NUCLEAR IMPL~~~~~ ~~

Nuclear Implosions: The Rise and Fall of the Washington Public Power Supply System follows a small public agency in Washington State that undertook one of the most ambitious construction projects in the nation in the 1970s: the building of five large nuclear power plants. By 1983, delays and cost overruns, along with slowed growth of electricity demand, led to cancellation of two plants and a construction halt on two others. Moreover, the agency defaulted on $2.25 billion of municipal bonds, leading to a monumental court case that took nearly a decade to resolve fully. Daniel Pope sets this in the context of the postwar boom's ending, the energy shocks of the 1970s, a new restraint in forecasting demand, and shifting patterns of municipal finance. *Nuclear Implosions* also traces the entangling alliance between civilian nuclear energy and nuclear weapons and recounts a telling example of how the law has become a primary method of resolving disputes in a litigious society.

Daniel Pope (Ph.D. Columbia University, 1973) is an American historian teaching at the University of Oregon since 1975. Pope is the author of *The Making of Modern Advertising* (1983) and many articles and reviews on the history of advertising, marketing, and consumer culture, and he is the editor of *American Radicalism* (2001). Pope was the Harvard-Newcomen Postdoctoral Fellow in Business History at Harvard Business School (1980–1981), held two Fulbright Senior Lecturer positions (University of Rome, 1996; Copenhagen Business School, 2004), and received the University of Oregon's Burlington-Northern Distinguished Teaching Award in 1989.

Nuclear Implosions

The Rise and Fall of the Washington Public Power Supply System

DANIEL POPE

University of Oregon

CAMBRIDGE
UNIVERSITY PRESS

CAMBRIDGE UNIVERSITY PRESS
Cambridge, New York, Melbourne, Madrid, Cape Town, Singapore,
São Paulo, Delhi, Dubai, Tokyo, Mexico City

Cambridge University Press
32 Avenue of the Americas, New York, NY 10013-2473, USA

www.cambridge.org
Information on this title: www.cambridge.org/9780521179744

First published 2008
First paperback edition 2010

A catalog record for this publication is available from the British Library

Library of Congress Cataloging in Publication data
Pope, Daniel, 1946–
Nuclear implosions : the rise and fall of the Washington public power supply system /
Daniel Pope.
p. cm.
Includes index.
ISBN-13: 978-0-521-40253-8 (hardback)
1. Washington Public Power Supply System.   2. Nuclear power plants – Washington
(State) – Design and construction – Finance.   3. Electric utilities – Washington (State) –
Finance.   4. Energy development – United States.   1. Title.
HD9685.U7W3456   2008
333.793′209797–dc22   2007050077

ISBN 978-0-521-40253-8 Hardback
ISBN 978-0-521-17974-4 Paperback

To my daughter, Stephanie C. Pope,
and the memory of my mother, Edith G. Pope

Contents

List of Tables

Acknowledgments

Writing history often seems to the scholar like a solitary venture, but in reality it is a collective enterprise. Over the many years that I have studied the Washington Public Power Supply System's nuclear projects, I have accumulated many debts. The Supply System's financial debts proved to be its undoing, but repaying intellectual debts by acknowledging the generosity of others is a pleasure.

Librarians, archivists, and other staff paved the way for research at the Washington Public Power Supply System (now Energy Northwest) itself, in Richland, Washington; the Eugene (Oregon) Water and Electric Board; the Springfield (Oregon) Utility Board; University of Washington Special Collections, Seattle; Washington State Archives (Olympia); Bonneville Power Administration Library (Portland, Oregon); and Seattle Municipal Archives. Special thanks go to Justice Martha Lee Walters of the Oregon Supreme Court, who, while in private practice, lent me documents accumulated in representing a utility in one of the cases I discuss. In conversations, Eve Vogel provided the perspectives of a talented geographer and reminded me of the economic, environmental, and cultural significance of salmon in the Northwest. Although I was reluctant for a long time to expose the messy construction site of a manuscript in progress to the eyes of critical readers, as I neared the end of the project I asked several colleagues and experts to read part or all of this study. In particular, my old friend David Lipton, Professor of Law at the Catholic University of America, read drafts of chapters 5 and 6 and did his best to explain to me the fundamentals of securities law and regulation. Professor John Findlay of the University of Washington and Professor Tom Wellock of Central Washington University read the whole manuscript. Each offered good advice on key historical issues. Professor Findlay and Professor Bruce Hevly (also of the University of Washington) convened a stimulating conference on "The Atomic West" in 1992. I presented a paper there; a later version, "Antinuclear Activism in the Pacific Northwest: WPPSS and Its Enemies," appeared in their edited volume, *The Atomic West* (Seattle and London: University of Washington Press, 1998). Professor Derek Hoff of Kansas State University, a former graduate student, more than repaid the value of advice I had offered him over the years by his close reading of several chapters. I benefited enormously from a detailed commentary by John Shurts, General Counsel of the Northwest Power and Conservation Council as well as an outstanding legal historian. John also provided an analysis of the debt repayment costs of the terminated nuclear projects that Pacific Northwest ratepayers still bear.

Several people involved in the Supply System's story answered my questions and gave me their judgments in interviews over the years. I am grateful for their cooperation and candor. They include Robert Ackerman, U.S. Representative Peter DeFazio, Robert Ferguson, Howard Gleckman, former Bonneville Power Administration head Peter T. Johnson, Jim Lazar, C. Richard Lehmann, Frank McElwee, Jane Novick, Keith Parks, Richard Quigley, and former U.S. Representative Jim Weaver.

At the University of Oregon, many people at the Knight Library provided valuable assistance. I mined several of its electronic resources extensively in this study. Interlibrary Loan brought me books and articles from afar. The Oregon Collection was a treasure trove of rarities and ephemera. Faculty colleagues in the History Department and around campus offered good judgment and advice as well as a lively intellectual atmosphere and fine examples of creative scholarship. Institutionally, a University of Oregon Summer Research Award and a grant from the University's Oregon Humanities Center helped in early stages of this research.

Frank Smith at Cambridge University Press found value in this project many years ago and maintained his support for it. Eric Crahan has been both professional and encouraging as the book has moved toward completion. Others at Cambridge have been helpful as well. Himanshu Abrol at Aptara has handled production matters skillfully and responsively.

My daughter, Stephanie Pope, helped me index and arrange clippings and lived a good deal of her adolescence with the early years of this study. As an adult, she has provided encouragement and motivation for me to complete it. My wife, Barbara Corrado Pope, read the entire manuscript twice. Her talents as an historian, writer, and editor have made this a better book in innumerable respects. Her uncanny ability to know when and how to prod me on may not be evident in the text, but without her advice and support, this book would be another terminated nuclear project.

Preface

On June 22, 2005, President George W. Bush journeyed to the Calvert Cliffs nuclear plant in southern Maryland, about fifty miles outside Washington, D.C. Speaking there, he proclaimed, "It is time for this country to start building nuclear power plants again." The president's endorsement came as no surprise. The press pointed out that he was the first president to come to a nuclear plant since Jimmy Carter had gone to Three Mile Island during the Pennsylvania reactor's 1979 crisis. Few newspapers reported two days later that President Carter himself visited the Cook Nuclear Plant in western Michigan and offered his own support for a revival of nuclear generation. "I think the future holds great opportunities for nuclear power," Carter stated.[1] Although Carter had served as a nuclear submarine engineer prior to his political career, he had been viewed as generally unsympathetic to nuclear power's growth during his term in office. Yet by 2005 a pro-nuclear political consensus seemed to be emerging.

Even before the two presidents spoke, there were many signs of revived interest in nuclear power. Utilities, reactor vendors, and construction firms had formed three consortia to explore potential projects in the United States. The Nuclear Regulatory Commission had, as far back as 1989, streamlined licensing procedures. The Energy Policy Act of 2005 contained several inducements to start a new round of nuclear construction. Rapidly rising oil and natural gas prices, along with projections of an

---

[1] For Bush's visit, see, e.g., Matthew L. Wald, "On a Rare Visit, Bush Talks Up Atomic Power," *New York Times*, June 23, 2005. His speech at the plant can be found at http://www.whitehouse.gov/news/releases/2005/06/20050622.html, accessed September 6, 2005. Carter's visit is described in "Former President Carter Highlights Nuclear Energy's Role During Tour of AEP's Cook Nuclear Plant," PR Newswire US, June 24, 2005, on Lexis-Nexis Academic, accessed September 2, 2005.

imminent peak in world petroleum output followed by a long, painful period of decline, persuaded analysts and policy makers to look afresh at non-petroleum energy sources. Even some environmentalists had come to view nuclear power as a preferable alternative to fossil fuels' carbon emissions and the climate change they cause.[2] Public opinion was also shifting in a positive direction. According to one survey, 70 percent of those polled indicated support for nuclear power in 2005, up from 46 percent a decade earlier. Communities under consideration as sites appeared receptive.[3]

Yet prospects for a full-fledged nuclear revival in the United States are cloudy. Construction of the long-delayed permanent waste repository planned for Yucca Mountain, Nevada, seems to recede endlessly into the future. The attacks of September 11, 2001, raised concerns about terrorist attacks on reactors and raised questions about the vulnerability of spent-fuel storage ponds at nuclear plant sites.[4] However, the greatest impediment to resuming nuclear construction in the United States is financial. Not only the foes of nuclear energy but also many of its corporate backers agree that investing in nuclear power at this point would be a risky venture. "Moody's would go bananas if we announced we were going to build a nuclear plant," said Thomas E. Capps. Capps is CEO of Dominion Resources Inc., a major electricity and natural gas supplier and the lead firm in one of the consortia investigating nuclear construction. Marilyn Kray, the president of NuStart Energy Development, the largest of the consortia, was more guarded: "There is much more confidence in the new process [of regulation and licensing], but not enough yet to make a new investment. Financiers are saying they are not yet comfortable."[5]

[2] "New Nuclear Wins Big in White House/Congress Energy Bill Deal," *Energy Washington Week*, August 3, 2005; Tom Ichniowski, "Energy Bill, Set to Be Signed, Is Filled with Industry Goodies," *Engineering News-Record*, August 8, 2005, p. 13. For a knowledgeable, if controversial, case that world oil output is peaking, see Kenneth Deffeyes, *Beyond Oil: The View from Hubbert's Peak* (New York: Hill and Wang, 2005). A useful overview of future energy options is Michael Parfit, "Future Power: Where Will the World Get Its Next Energy Fix?", *National Geographic*, v. 208, n. 2, pp. 2ff.; on environmentalist responses, see Felicity Barringer, "Old Foes Soften to New Reactors," *New York Times*, May 15, 2005, p. 1.

[3] John Carey, "Maybe in My Backyard," *Business Week*, September 5, 2005, p. 68.

[4] On the threats to spent-fuel ponds, see Matthew L. Wald, "Agencies Fight over Report on Sensitive Atomic Wastes," *New York Times*, March 29, 2005, p. 14.

[5] Shankar Vedantam, "Uncertainties Slow Push for Nuclear Plants," *Washington Post*, July 24, 2005, p. A6; Ralph Vartabedian, "Nuclear Industry Lays Foundation for Comeback," *Los Angeles Times*, June 22, 2005. Other examples of doubts within the industry can be found in Melissa Leonard, "Will U.S. Nuclear Power's 'Renaissance' Have a Short Half-Life?", *Power*, v. 149, n. 4 (May 2005), pp. 34–35.

The story of the Washington Public Power Supply System's attempts a generation ago to build five large nuclear power plants should give utilities and financiers further reason to proceed with great caution. The results of the Supply System's efforts are simple to recount: one plant completed, two terminated in 1982, and another two canceled in 1994, after more than a decade-long construction moratorium. Shortly before the 1982 terminations, the agency estimated costs for building the five plants at $23.9 billion, more than five times the total of the projects' initial estimates. In 1983, legal difficulties forced the Supply System to default on $2.24 billion in municipal bonds that it had sold; this, in turn, led to a securities fraud suit of enormous magnitude and Dickensian complexity.

The lessons of the Washington Public Power Supply System debacle are ambiguous. Nuclear proponents can rightly point out that the Supply System's organizational failings were unusual if not unique, even in the troubled history of American nuclear projects. The court case that brought on the 1983 default was surely a legal anomaly for the nuclear power industry. The political climate for nuclear energy, especially in the aftermath of Three Mile Island, was substantially less welcoming than it is today. The stagflation of the 1970s and the sharp recession of 1981–1982 posed daunting problems for finance. In 2005, only a Pollyanna would state that such conditions could never be repeated; only a Cassandra would claim they were inevitable.

Nevertheless, although circumstances and protagonists have changed since the Supply System's undertakings, an understanding of what happened in the Pacific Northwest a quarter-century and more ago should prove illuminating. Will policymakers turn unquestioningly, as they did in that era, to supply-side solutions for electrical energy? In a competitive utility environment, will utilities and other players strive to build organizational empires without the resources to succeed or the judgment to know whether organizational growth will solve problems or create new ones? A generation ago, WPPSS, the Bonneville Power Administration, and others in the utility community professed a democratic ethos but often reacted to popular pressures with hostility. Can institutions be open and responsive to citizen, consumer, and environmentalist concerns? When conflicts arise over complex technical, legal, and economic issues, can the judicial system resolve them acceptably? In the past, civilian nuclear energy was intertwined with the Cold War and nuclear weaponry. Will our energy policies in decades to come be calibrated with military ambitions?

The Washington Public Power Supply System's story is not simply about nuclear power. It touches on some of the most important developments in contemporary America's political economy: the shift from buoyant expectations of growth to an awareness (though not always an acceptance) of limits; the intermingling of military and civil institutions; the complexities of prediction and planning in an era of large-scale institutions and undertakings; the problems that arise when technological possibilities outrun organizational capacities for large-scale projects; new forms of environmental and consumer activism; and the costs of making decisions and resolving disputes in a litigious society.

Within the framework of a chronological narration, *Nuclear Implosions* seeks to bring historical insights to bear on these vital questions. Chapter 1 traces the Pacific Northwest's distinctive commitment to electrical energy as the key to economic growth. The great federal dams on the Columbia River and the central position of the Bonneville Power Administration meant that public power was stronger and the battles between public and private utilities fiercer in this region than in most of the rest of the country. Households and businesses thrived at mid-century on low-cost hydropower and turned to nuclear energy as the next choice when hydro capacity approached its limits. Officials of the Washington Public Power Supply System, a consortium of local public utilities in the state of Washington, thought of the agency as the rightful heir to the progressive public power movement in the Northwest. They also considered it their task to facilitate growth, which they assumed would come in tandem with an expanding electrical supply.

In chapter 2, we encounter Washington Public Power Supply System's commitment to build three large nuclear power plants. Like other utilities embarking on nuclear projects in the 1960s and 1970s, the Supply System employed reactor designs based on the devices that propelled nuclear submarines. Here, and throughout the narrative, we see the Supply System's nuclear efforts closely meshed with the military nuclear projects of Cold War America. Notably, the organization's nuclear baptism came with its designation to build and operate turbine-generators attached to a facility producing weapons plutonium on the Hanford Military Reservation. Its headquarters came to be sited only a few hundred yards from the entrance to the Reservation, and three of its nuclear plants were to be built on the Reservation.

Chapter 3 describes the process in the mid-1970s that drew the Supply System and eighty-eight public utility participants into undertaking two additional plants. Projects 4 and 5 lacked the implicit financial guarantee that the Bonneville Power Administration had provided for the first three plants. Local utilities, torn between financial caution and alarming predictions of impending power shortages, took shares in the new projects' "capabilities." Their experience bears out Niels Bohr's aphorism, "Prediction is very difficult, especially about the future." Cost estimates soared, estimated completion dates receded into the future, and electricity demand fell far short of predictions. Indeed, "Long-term forecasts of energy affairs...have...a manifest history of failure," as Vaclav Smil, a leading scholar in the field, has put it.[6] The Supply System's experience demonstrates that failure vividly. Meanwhile, the pressures that induced the Supply System to take on these projects suggest that organizations that claim to represent the people's interests and desires find themselves pushed and pulled by political and economic forces remote from, if not opposed to, the popular will.

Chapter 4 shows the chasm between the enormous tasks that the Supply System had assumed and the resources and capabilities of the organization. The chapter draws on insights from sociological theories of organization and management studies, especially in the field of construction management. They help us not only to understand why the Supply System's projects went so badly off course but also to see ways that these failures paralleled those in other large nuclear construction efforts and, in fact, a wide variety of other types of construction, transportation, and infrastructure projects.

Chapter 5 focuses on termination of Projects 4 and 5 in 1982 and the legal and financial imbroglio that caused the Supply System's gargantuan bond default the next year. The proximate cause of the default was a controversial 1983 court ruling that the participating utilities had lacked authority to contract for capability shares seven years earlier and therefore were not obligated to pay back the projects' bondholders. The broader context of termination and default included a severe and ominous recession, sharply slower growth of electrical demand, ratepayer fury at the prospect of paying billions for power that would never be delivered, and a lightly regulated municipal bond market offering unprecedented high interest rates for investments that turned out to be very risky.

In the aftermath of default, the locus of the Washington Public Power Supply System's history shifted, in large measure, from construction sites to courtrooms. Chapter 6 describes this. One of the largest class-action

---

[6] Vaclav Smil, *Energy at the Crossroads* (Cambridge, Mass. and London: MIT Press, 2003), 121.

lawsuits in recent history pitted about 75,000 aggrieved bondholders against hundreds of individual and organizational defendants. Beyond the recondite particulars of securities law, fundamental issues about financial risk and legal responsibility were at stake. The case also raised serious questions about the legal system's ability to resolve controversies of this magnitude and complexity. For the future, the case stimulated reforms in the municipal bond market, but whether these were sufficient to ward off major problems is very much open to doubt.

An epilogue (chapter 7) examines major developments on the electrical energy scene since the 1980s and reflects on the implications of the Supply System story for broader issues. Much has been changed – including the Supply System itself, now renamed Energy Northwest and engaged in an expanded mission. However, basic questions remain unresolved. Have we learned anything about forecasting the future from the mistakes of the 1960s and 1970s? In providing electricity, will deregulation and competition replace the pattern of regulated private utilities with a substantial minority of public systems that dominated the twentieth century? More generally, how do we balance environmental, security, and economic concerns in energy policy? What role will nuclear energy play? This book tells a story of misguided planning that exacted a high financial cost. Now mistakes and shortsighted energy policies will cause far greater harm, to the ecosystem and to humanity. The stakes are higher than ever.

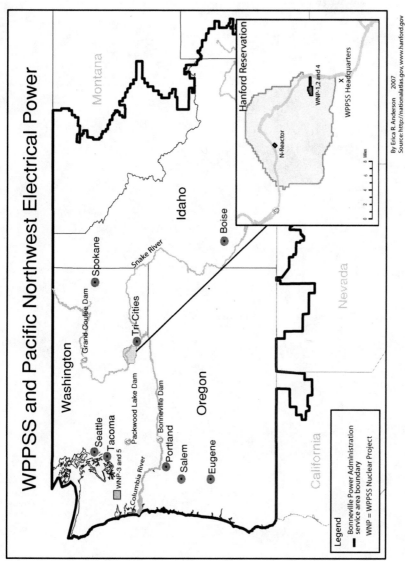

WPPSS and Pacific Northwest Electrical Power

Montana

Washington

Grand Coulee Dam

Spokane

Snake River

Tri-Cities

Seattle
Tacoma
WNP-3 and 5

Packwood Lake Dam

Bonneville Dam

Columbia River

Portland

Salem

Oregon

Eugene

Idaho

Boise

Nevada

California

Legend
Bonneville Power Administration
service area boundary
WNP = WPPSS Nuclear Project

Hanford Reservation

N-Reactor

WNP-1, 2 and 4

WPPSS Headquarters

0 2 4 6 8
Miles

By Erica R. Anderson   2007
Source: http://nationalatlas.gov,
www.hanford.gov

WPPSS and Pacific Northwest Electrical Power by Erica R. Anderson 2007. *Source:* http://nationalatlas.gov,
www.hanford.gov

I

Background to Fiasco

In June 1968, while many communities in the Pacific Northwest were preparing to celebrate their centennials, Richland, a small city in southeastern Washington, commemorated its tenth anniversary as an incorporated city and the twenty-fifth anniversary of the community's modern beginning, the designation of Hanford, Washington, as the site of plutonium production for the Manhattan Project. Amid the reminiscences and self-congratulations of the 1968 festivities, one highlight was Glenn Seaborg's banquet address on Friday, June 7.

Seaborg was truly one of the high priests of the nuclear era. While still in his twenties, he had been a co-discoverer of plutonium. In early 1942, he launched an extensive research program to isolate the element in quantities sufficient for bomb production. His success paved the way for Hanford's mission in the Manhattan Project, manufacturing enough plutonium for the "Trinity" bomb exploded above New Mexico and the "Little Boy" weapon dropped on Nagasaki on August 9, 1945.

When Seaborg visited Richland and the Hanford site in 1968, he was chairman of the Atomic Energy Commission. His address to the dignitaries that evening presented his vision of a peaceful nuclear America. Hanford had just been chosen to house the Fast Flux Test Facility, a breeder reactor development project. For Seaborg, this was only the start. In the future, a complex of very large breeder reactors could generate vast quantities of cheap electricity and industrial process heat. Heavy industry could locate in the complex – a "nuplex." In the words of the local newspaper, "With the nuplex, conventional resources could be processed more cheaply; new and exotic materials would be produced, and most of the waste could be recycled on an economic basis." Seaborg held out promises of a "junkless society" and of heavy industry separated from major cities, which would, "once again become a place primarily for people." He reflected, "Perhaps 25 years from now we will be able to gather here to look back over a half a century of progress of the nuclear age.... And we will be able to reminisce about the beginning of the nuclear age while we see about us many of the wonders that it has brought and continues to unfold."[1]

More than sixty years after the Hanford facility was built, the nation's nuclear fate must give pause to the followers of Seaborg's dream. The Hanford Generating Project, which used steam from a plutonium-producing reactor to produce electricity, closed down in the wake of the Chernobyl disaster of 1986. There is no American breeder reactor. Congress killed funding for the Clinch River (Tennessee) reactor, the centerpiece of the breeder program, in 1983. Energy Secretary James Watkins placed Hanford's Fast Flux Test Facility, which originally was to irradiate fuel for Clinch River, on "cold standby" in the waning days of the first Bush administration.[2] In 2006, as work on a permanent shutdown was underway, the American Nuclear Society designated the reactor a National Nuclear Historic Landmark. There is no nuplex, at Hanford or elsewhere, no cornucopia of costless energy and clean manufacturing. Indeed, no utility in the country has ordered a nuclear reactor since 1978. Once expected to serve half of the nation's electricity needs by the year 2000, nuclear reactors generate only about one-fifth.

Commissioner Seaborg's predictions for Hanford and the Pacific Northwest were far off the mark. Following the shutdown of the Hanford Generating Project, Portland General Electric Company, owner of the Trojan nuclear plant in Rainier, Oregon, closed it in 1993, leaving only

---

[1] Glenn T. Seaborg, "Large-Scale Alchemy – 25th Anniversary at Hanford-Richland," in Seaborg, *Nuclear Milestones* (San Francisco: W. H. Freeman, 1972), 162–175. Nuplex reference is at p. 175. "Richland Is Human Bonus, Spin-Off of Nuclear Efforts, Says Seaborg," *Tri-City Herald*, June 9, 1968. These were not Seaborg's first utopian speculations about the nuclear age. In 1948, he predicted that "the atomic energy unit will sprout great wings and take to the upper air," propelling commercial airliners. Cited in Stephen L. Del Sesto, "Wasn't the Future of Nuclear Energy Wonderful?" in Joseph J. Corn, ed., *Imagining Tomorrow: History, Technology, and the American Future* (Cambridge, MA: M.I.T. Press, 1986), 65. As late as 1980, a researcher told a Washington State legislative committee that Hanford could become an "energy park" with as many as twenty nuclear reactors at the site. "Hanford pushed as energy park," *Tri-City Herald*, February 10, 1980. Indeed, in 2001, former Congressman Sid Morrison stated his desire to see part of Hanford used as an energy park – Chris Mulick, "Lawmaker named to Energy Northwest executive board," *Tri-City Herald*, July 26, 2001.

[2] "Status Lowered Again: Hot Standby to Cold," *Nuclear News*, March, 1993, 87. See also William Walker, "The Back-End of the Nuclear Fuel Cycle," *History and Technology*, 9 (1992): 189–201.

one functioning power reactor in the region, the Washington Public Power Supply System's (now Energy Northwest's) Nuclear Plant 2 (now renamed the Columbia Generating Station) at Hanford. This facility represents only a small fraction of the Supply System's grandiose plans. It had committed to build and operate five large plants. Yet today hydropower remains the region's basic source of electricity; regional energy planners contemplate conservation, renewable resources, and small-scale natural gas–generating facilities rather than large nuclear reactors. In a disturbingly ironic turn, the area around Hanford itself is now a focus of environmental anxiety for the Northwest. Airborne radiation releases, both unplanned and deliberate, liquid wastes in aging single-walled underground tanks, and lapses in worker and community health procedures all make cleaning up the primary mission of Hanford in a post–Cold War era. The nuclear alchemists' handiwork at Hanford and elsewhere is now our problem and our descendants' burden.[3]

The 1968 Hanford observances took place against a backdrop of intensive planning to bring nuclear energy to the Northwest. Four months later, utility leaders announced a Hydro-Thermal Power Program.[4] As the name suggested, the plan proposed a shift from almost exclusive reliance on energy from water spilling over the giant dams on the Columbia River and its tributaries to a hybrid system where thermal* generation (steam driving turbines) would supply the base load for residential, agricultural, commercial, and industrial customers. Hydropower would serve users in times of peak demand. Nothing if not ambitious, the Hydro-Thermal

---

[3] Michele Stenehjem Gerber, *On the Home Front: The Cold War Legacy of the Hanford Nuclear Site* (Lincoln and London: University of Nebraska Press, 1992) is a major treatment of the environmental impact of military activity at Hanford; Patricia Nelson Limerick, "The Significance of Hanford in American History," in David H. Stratton, ed., *Washington Comes of Age: The State in the National Experience* (Pullman: Washington State University Press, 1992), 153–171, offers an overview of Hanford's role. See also several revealing articles by Karen Dorn Steele in the *Bulletin of Atomic Scientists*: "Hanford's Bitter Legacy," 44, 1 (January/February 1988): 17–23; "Hanford: America's Nuclear Graveyard," 45, 8 (October 1989): 15–23; "National Security Ever Green," 45, 10 (December 1989): 6; "Tracking Down Hanford's Victims," 46, 9 (October 1990): 7–8, 46.

[4] U.S. Department of the Interior, Bonneville Power Administration, *A Ten Year Hydro-Thermal Power Program for the Pacific Northwest* (January 1969). The program was first publicized in October 1968; the published document is dated January 1969.

* Some definitions: Thermal generation uses heat to make steam to drive turbines. Fossil fuel generation is a subset, in which the fuel is, in almost all cases, coal, natural gas, or petroleum. Nuclear reactors create thermal power, but nuclear materials are not fossil fuels.

Power Program called upon utilities in the region to build and operate ten large thermal plants by 1980. Extrapolating further, it suggested that the Northwest would need another ten generating facilities in the following decade. These projects would supply a demand expected to triple by 1990.

Even before the Hydro-Thermal Power Program, the Washington Public Power Supply System had volunteered to undertake a large nuclear plant for the consumer-owned utilities of the region. At the time, the Washington Public Power Supply System was a small organization headquartered in Kennewick, adjacent to Richland. By 1976, this agency had agreed to finance, build, and operate five large nuclear power plants to help meet the region's predicted energy needs. By the end of the 1970s, WPPSS* had become the largest single municipal borrower in the nation. Fourteen thousand workers were building the plants at the peak of construction activity in 1981. The Supply System itself, which had 81 employees when construction began in 1971, employed a staff of over 2,000 a decade later.[5]

But gargantuan plans meant colossal problems. The plants all fell far behind schedule and costs soared. Demand for electricity lagged well behind earlier predictions. In 1980, the Supply System Board of Directors forced the Managing Director to resign and hired Robert L. Ferguson, who had been Deputy Assistant Secretary of Energy in the Carter administration. Ferguson soon called for a thorough budget review. In May 1981, the study disclosed that the estimated total cost of the five plants would be $23.9 billion, more than five times original estimates. Ferguson slowed down construction on Plants 4 and 5. By the following January, he felt forced to call for terminating these projects. Three months later, the Supply System imposed a construction moratorium on Plant 1, and in 1983 halted work on Plant 3. When Plant 2 finally opened in December 1984, the region already had a surplus of electric power, a condition that lasted into the next decade. Table 1.1 summarizes the projects' ownership, financing, costs, and eventual fate.

---

* Because the initials WPPSS would be pronounced as "Whoops," the Washington Public Power Supply System tried to avoid this abbreviation. Its managers referred to the agency as the Supply System. This book will use the two shorthand versions interchangeably, with no invidious connotations meant to apply to WPPSS. In 1998, the Supply System changed its name to Energy Northwest. In references to events from that year on, we will call the organization Energy Northwest.

5 James Leigland and Robert Lamb, *WPPSS: Who Is to Blame for the WPPSS Disaster* (Cambridge, MA: Ballinger, 1986), 24.

TABLE 1.1. *WPPSS nuclear projects summary*

	WNP-1	WNP-2	WNP-3	WNP-4	WNP-5
Location	Hanford Reservation	Hanford Reservation	Satsop	Hanford Reservation	Satsop
Ownership	public	public	70% public/30% private	public	90% public/10% private
Financing	net billed	net billed	net billed	88 participating utilities' shares of "project capability"	88 participating utilities' shares of "project capability"
Initial Expected Completion Date	Sept. 1980	Sept. 1977	March 1982	March 1983	Jan. 1985
Initial Financing Estimate and Estimate Date*	$1.204 billion Sept. 1975	$507 million June 1973	$993 million Dec. 1975	$3.377 billion† Feb. 1977	
Cost Estimate May 1981‡	$4.268 billion	$3.216 billion	$4.532 billion	$5.510 billion	$6.261 billion
Final Status	Construction halted April 1982; project terminated May 1994	Project completed December 1984	Construction halted May 1983; project terminated May 1994	Construction halted May 1981; projects terminated Jan. 1982	Construction halted May 1981; projects terminated Jan. 1982

* Initial financing estimates are the amounts announced at the time of the initial borrowing for each project.

† WNP-4 and WNP-5 were financed jointly through bond issues for both projects. Bonds were issued for each net-billed plant separately.

‡ *Source:* Washington Public Power Supply System, *Annual Report*, 1981, Financial Section, 23–24.

Source: Memo, Thomas P. Friery, Manager, Treasury Division, WPPSS, to Peter J. D. Gordon, T. Rowe Price Associates, Inc., July 10, 1978, reprinted in U.S. House of Representatives, Committee on Interior and Insular Affairs, Subcommittee on Mining, Forest Management and Bonneville Power Administration, *The Bonneville Power Administration [BPA] and Washington Public Power Supply System [WPPSS]*, Serial No. 98–48, Part I (Washington: U.S. Government Printing Office, 1985), 87.

Meanwhile, the termination of Projects 4 and 5 set in motion a legal struggle over the $2.25 billion that WPPSS had borrowed for these facilities. Eighty-eight Northwest public utility districts, municipal utilities, and rural electrical cooperatives had signed Participants' Agreements in 1976 that seemed to bind them to pay for shares of the projects' generating capabilities, whether or not they were successfully completed. In June 1983, however, the Washington State Supreme Court ruled that utilities in that state had lacked authority to enter into the Participants' Agreements. Hence, the court argued, the utilities did not have to make payments on the bonds. Northwest political and business leaders scrambled for other ways to meet the bond obligations, but in August 1983, WPPSS defaulted. In a sad addition to the Supply System's list of superlatives, this was the largest municipal bond default in American history.

This book describes the rise and fall of the Supply System. It is a story rooted in the Pacific Northwest's distinct regional history. The developmental role of low-cost hydroelectric power, exemplified by the Bonneville and Grand Coulee Dams on the Columbia River and the establishment of the Bonneville Power Administration in 1937, made Northwesterners acutely aware of the importance of electricity to the region's economy. With the exception of the Tennessee River Valley, nowhere else did the federal government play such a large role in electrical energy generation and transmission. Moreover, nowhere else in early and mid-twentieth century America did the politics of electricity stay on center stage for so long. Struggles between public and private power interests were features of the region's political climate for decades. Some of the Supply System's leaders and cheerleaders saw the agency as the heir of the Northwest's public power pioneers.

Yet the Washington Public Power Supply System did not stand in isolation in the "far corner" of the contiguous United States. It reflects several nationwide trends in energy policy and public utilities. The WPPSS collapse is the story of fatally flawed demand projections, incessant problems of construction management, and thorny political conflicts on the uneasy borderline between public and private sectors. Accounts of nuclear energy projects across the country reveal similar problems.[6]

[6] Henry F. Bedford, *Seabrook Station: Citizen Politics and Nuclear Power* (Amherst: University of Massachusetts Press, 1990); David P. McCaffrey, *The Politics of Nuclear Power: A History of the Shoreham Nuclear Power Plant* (Dordrecht, Netherlands: Kluwer Academic Publishers, 1991); Joseph P. Tomain, *Nuclear Power Transformation* (Bloomington: Indiana University Press, 1987).

This is also a story about some of the most crucial developments in the recent American political economy: the shift from an era of buoyant expectations of growth to an age of limits; the emergence of new, high-stakes, and risky ventures in finance; the new environmental and consumer movements of the late twentieth century; the costs of making decisions and resolving disputes in a litigious society; the problems that arise when technological possibilities outrun organizational capacities for large-scale projects. Finally, as earlier commentators on the history of nuclear energy have observed, civilian nuclear power bears a close though ambiguous relationship to the military uses of the atom.[7] With three of its reactors sited on the Hanford Nuclear Reservation, WPPSS indicates some of the subtle yet important ways that war, hot and cold, has permeated American society.

The Supply System's ventures, therefore, need to be situated in both their national and their regional contexts. With over thirty years of hindsight, the decisions to move toward a nuclear electric energy base in the Northwest present a picture spotted with folly and ineptitude. Looking at the lessons that generation of power planners drew from prior experience will help to make their choices understandable even if we cannot deem them wise.

The Hydroelectric Legacy

The Columbia River has for centuries been the source of much of what was and is distinctive about the Pacific Northwest.[8] Native Americans

---

[7] Important works on nuclear power in the United States include Mark Hertsgaard, *Nuclear Inc.* (New York: Pantheon, 1983); Brian Balogh, *Chain Reaction: Expert Debate and Public Participation in American Commercial Nuclear Power, 1945–1975* (Cambridge and New York: Cambridge University Press, 1991); Joseph G. Morone and Edward J. Woodhouse, *The Demise of Nuclear Energy?* (New Haven and London: Yale University Press, 1989); Spencer R. Weart, *Nuclear Fear: A History of Images* (Cambridge, MA and London: Harvard University Press, 1988); John L. Campbell, *Collapse of an Industry* (Ithaca, NY and London: Cornell University Press, 1988).

[8] Michael S. Spranger, "The Columbia River: The Pacific Northwest's Most Precious Resource," *Pacific Northwest Forum*, 9 (Summer/Fall 1984): 3–16 provides a useful introduction to the Columbia's history. Many more recent studies emphasize the salmon crisis that emerged in the 1990s. They include: Joseph Cone, *A Common Fate: Endangered Salmon and the People of the Pacific Northwest* (Corvallis: Oregon State University Press, 1996); Cone and Sandy Ridlington, eds., *The Northwest Salmon Crisis: A Documentary History* (Corvallis: Oregon State University Press, 1996); Blaine Harden, *A River Lost* (New York: W.W. Norton, 1996); William Dietrich, *Northwest Passage: The Great Columbia River* (New York: Simon & Schuster, 1995); Richard White, *The Organic*

have lived near its banks for at least twelve millennia, and coastal Indians traveled upstream to meet with inland peoples. The salmon that spawn upstream in the river and its tributaries remain an integral part of Native cultures, despite threats to the continued viability of the salmon runs. In the nineteenth century, the Columbia was the axis of transportation for European-American exploration, settlement, and commerce in the Oregon Territory. In the twentieth century, damming the Columbia River system provided the electric power that transformed the region.

The Columbia River is 1,243 miles long, slightly more than half the length of the Mississippi. From its source in the mountains of eastern British Columbia, it flows south, cutting through the arid territory of eastern Washington. The Snake River joins it near the Tri-Cities of Pasco, Kennewick, and Richland. Soon it turns west and forms the border between Washington and Oregon. At Portland, the Willamette River merges with it from the south. The Columbia flows into the Pacific near Astoria, Oregon. In its journey, the Columbia drops 2,650 feet in elevation, almost twice as large a decline as the Mississippi. Stream flow is prodigious, averaging 265,000 cubic feet per second. The water that flows through the Pacific Northwest's rivers almost equals the volume in all other rivers west of the Mississippi.

The laws of physics decree that the potential for generating electricity from falling water depends on these last two factors, the vertical fall and the volume of water. On both these counts, the Columbia and its tributaries served the region's electrical energy needs well. Indeed, the region has about one-third of the nation's hydroelectric capacity potential. From early in the twentieth century, there were those who pressed for hydropower development. Rufus Woods, owner and publisher of central Washington's *Wenatchee Daily World*, campaigned tirelessly for damming the Columbia and irrigating the Columbia Basin desert. J. D. Ross espoused the cause of public ownership and development of hydro resources as head of Seattle City Light, the municipal system. The Oregon and Washington Granges ardently advocated public power and pressed for legislation in both states that allowed formation of public utility districts.

Machine (New York: Hill & Wang, 1995); Jim Lichatowich, *Salmon Without Rivers* (Washington, D.C.: Island Press, 1999); Joseph E. Taylor, *Making Salmon: An Environmental History of the Northwest Fisheries Crisis* (Seattle: University of Washington Press, 1999); William L. Lang and Robert C. Carriker, eds., *Great River of the West: Essays on the Columbia River* (Seattle: University of Washington Press, 1999).

Yet in this sparsely populated region, the river system's capacity went untapped until the New Deal years. Early in the century, private power companies balked at the capital investment needed to bring electricity to rural areas, and conservatives contended that schemes to dam the river for power would end up lighting the desert for jackrabbits and rattlesnakes.[9] Franklin D. Roosevelt, however, had met James O'Sullivan, a reclamation engineer and development enthusiast, while campaigning for vice-president in Spokane in 1920. O'Sullivan persuaded Roosevelt, and the future president never changed his mind.[10] Campaigning in 1932, Roosevelt told an enthusiastic crowd in Portland that the federal government should exploit water power sites for the public benefit. These sources could provide a yardstick both to measure and to control the cost of private power.

FDR's pledge soon translated into action, and construction began on both Bonneville and Grand Coulee Dams in 1933, although Grand Coulee's official authorization did not come until passage of the 1935 Rivers and Harbors Act. Bonneville, located forty miles east of Portland, became an Army Corps of Engineers project, while the Bureau of Reclamation took charge of Grand Coulee, in north-central Washington.

Bonneville Dam was itself a formidable undertaking. The Columbia's banks at Bonneville were soft, not hard rock, and anchoring the dam was a struggle. The river's flow also complicated construction. Northwest rivers swell each spring as the mountain snowpack melts; in 1936, the waters broke through an earthen cofferdam above the Bonneville construction site and smashed into the partially completed structure. Yet, working around the clock, builders completed the dam on schedule, and President Roosevelt dedicated it on September 28, 1937. "We can well visualize a date, not far distant, when every community in this area will be wholly electrified," he proclaimed.[11]

Two months earlier, Roosevelt had signed the Bonneville Project Act to provide an administrative structure and policies for Northwest electric power. Leaders at the time viewed this as an interim measure to

---

[9] Wesley Arden Dick, "When Dams Weren't Damned: The Public Power Crusade and Visions of the Good Life in the Pacific Northwest in the 1930s," *Environmental Review*, 13 (Fall/Winter 1989):125–126 for critical comments about Grand Coulee Dam.

[10] John Gunther, *Inside U.S.A.* (New York: Harper & Brothers, 1947), 125; Herman C. Voeltz, "Genesis and Development of a Regional Power Agency in the Pacific Northwest, 1933–1943," *Pacific Northwest Quarterly*, 53 (April 1962): 65–76.

[11] Cited in Gene Tollefson, *BPA and the Struggle for Power at Cost* (Portland, OR: Bonneville Power Administration, 1987), 118.

neville Dam to function while politicians worked out a plan
e comprehensive Columbia Valley Authority modeled on the
Valley Authority. But repeated proposals for broad regional
development planning met defeat.[12] Nearly seventy years later, the act
and the agency it created, the Bonneville Power Administration (BPA),
continue to shape many of the Pacific Northwest's energy policies.

A key element of the enabling legislation resolved an intense bureaucratic feud and set regional power policy on a pro-development path.
Having built Bonneville Dam, the Corps of Engineers was eager not only
to operate it but to distribute its power. Yet the Corps believed that the
Northwest's market was limited. It proposed building only two short
transmission lines and giving industries near the dam site reduced rates.
That perspective appealed to Portland area businesses but angered public
power advocates in the rest of the region. These forces favored civilian control, an extensive transmission network, and uniform power rates
around the region, "postage stamp" pricing as it came to be called. In
early 1937, Washington Senator Homer T. Bone proposed a formula giving control of the dam to the Corps but establishing a civilian transmission
agency.

Placing BPA within the Interior Department put it under Secretary
Harold Ickes, a public power backer. The Bone compromise, by creating
a civilian agency in Ickes' department, appeared to benefit those favoring
comprehensive development and public power. Indeed, Roosevelt's choice
of J. D. Ross, perhaps the leading figure in the Northwest's public power
movement, as the first Administrator acknowledged this orientation. Ross
had an expansive view of Bonneville's mission. "It is not just what the electricity costs; it is what our people can do with it that constitutes the help
to humanity and makes it a real success."[13]

The Bonneville Project Act gave BPA a green light to become a regional
power transmission and marketing agency, the main role that it still plays
today. (In 1940, an Executive Order gave BPA the task of transmitting and
marketing power from Grand Coulee as well as Bonneville.) Conversely,
however, it established the principle that Bonneville cannot generate

---

[12] Charles McKinley, *Uncle Sam in the Pacific Northwest* (Berkeley and Los Angeles: University of California Press, 1952), 543–617 discusses Columbia Valley Authority proposals
and alternatives. For the complex politics of regional "little TVA" plans, see William
Leuchtenburg, "Roosevelt, Norris and the 'Seven Little TVAs'," *Journal of Politics*, 14,
3 (August 1952): 418–441.
[13] Philip J. Funigiello, *Toward a National Power Policy: The New Deal and the Electric
Utility Industry, 1933–1941* (Pittsburgh: University of Pittsburgh Press, 1973), 202.

electricity itself or own power-generating facilities. Nor can it distribute current at retail to individual customers, although it can sell electricity directly to a group of large industrial users, mostly aluminum companies, known as Direct Service Industries (DSIs).

Another central feature of the Bonneville Project Act was that public utilities (public utility districts, municipal systems, and electrical cooperatives) would receive preferential rights to the power BPA marketed. Section 4(a) of the Act read:

In order to insure that the facilities for the generation of electric energy at the Bonneville project shall be operated for the benefit of the general public, and particularly of domestic and rural consumers, the administrator shall at all times, in disposing of electric energy generated at said project, give preference and priority to public bodies and cooperatives.[14]

Reinforcing earlier federal pronouncements that hydroelectric development was meant for public utility systems and their customers, the "preference clause" has been a touchstone for public power advocates and a sore spot for privately owned utilities and industrial customers of the BPA. Public power advocates have been fond of citing remote precedents for preference, even finding support for the doctrine in Roman aqueduct development and public water-powered grist mills in the Plymouth Colony.[15]

In reality, public power preference is no eternal verity but a political choice subject to constant debate and redefinition. Until the 1990s, the federal hydropower system had been the cheapest electricity source in the Pacific Northwest (rivaled nationally only by Tennessee Valley Authority power). In part this resulted from the technological efficiencies of the hydro system. Additionally, the federal government has allowed Bonneville to pay for the federal investment in the system at interest rates below market levels. First call on these hydro resources historically allowed preference customers to charge rates among the lowest in the country. Within the Northwest, the differences between public and private utilities' rates were striking. In the late 1970s, for example, the cost of 1,000 kilowatt-hours to a customer of the Clark County (Washington) Public Utility District, in the Portland, Oregon, metropolitan area, was

[14] U.S. Code 16, Sec. 832c (a), also quoted in [Gus Norwood], *Columbia River Power for the People: A History of Policies of the Bonneville Power Administration* (Portland, OR: Bonneville Power Administration, n.d.), 69.
[15] Paul Nelson, "The Preference Clause – A Democratic Principle," *Pacific Northwest Public Power Bulletin*, 14, 2 (February 1960): 6.

$11.10. Across the river, buying the same amount of energy from one of the private companies serving the city of Portland cost a resident $27.00.[16]

Not surprisingly, while the preference customers fought tenaciously to maintain the policy, investor-owned utilities tried to change it. They maintained that the intended beneficiaries of preference were the "domestic and rural customers" of utilities in the region, not just the "public bodies and cooperatives" who served some but not all of these consumers. Since public power serves the majority of Washington State but only a minority of consumers in Oregon, Idaho, and Montana, the public preference issue divided the region geographically.

From the start, the Bonneville Power Administration had to maneuver its way through the dangerous currents of regional power politics. Public ownership interests hoped that J. D. Ross and his successor, Paul Raver, would use the organization's resources to encourage formation of public utilities and to help these utilities acquire resources from the private companies, either through negotiations or condemnation proceedings. However, more crucial to Bonneville's survival was the need to find markets for the energy the dams generated. Power sales to private utilities angered some public power proponents, but, Ross and his allies maintained, these were necessary to generate revenue while other markets emerged. Building transmission lines through the region and connecting Bonneville with Grand Coulee were key strategies for building demand.

From the moment of its arrival, federal hydropower in the Northwest brought to the public ownership movement a powerful impetus for growth and development. In Washington, fifteen Public Utility Districts (PUDs) won voter approval in 1936 in anticipation of Bonneville's completion. In elections in 1938 through 1940, eleven more were formed. Because of Oregon's more restrictive law, and a generally more conservative political climate, public power grew more slowly in that state. Even today, the majority of Washington's electricity comes from public utilities; power in Oregon (and Idaho) remains predominantly in the hands of private firms. Meanwhile, encouraged by New Deal legislation, rural electric cooperatives started up throughout the Northwest. By 1940, there were sixty-three new consumer-owned utilities.[17]

The scale of Grand Coulee Dam dwarfed Bonneville Dam. Its twelve million cubic yards of concrete make it the largest concrete structure in

---

[16] Kai N. Lee and Donna Lee Klemka with Marion E. Marts, *Electric Power and the Future of the Pacific Northwest* (Seattle: University of Washington Press, 1980), 29

[17] Tollefson, *BPA and the Struggle,* 185.

the United States and one of the largest in the world. It is the biggest single producer of electricity in this country, with a capacity almost twice as great as the runner-up.[18] With Grand Coulee's completion expected in 1941, electric supply was likely to leap ahead of demand. It took World War II to create a need for electricity to match the generating capacity of the Columbia dams. Alcoa announced plans for the first aluminum reduction plant in the region in December 1939. Bonneville eagerly agreed to build a substation and power line in order to serve the mill directly. Four more plants had opened by the end of the war, three of them built by the federal Defense Plant Corporation. These were voracious consumers of power and throughout the century continued to buy huge amounts of electricity from Bonneville. (In 1976, as WPPSS expanded to five nuclear projects, over 40 percent of Bonneville's firm power and nearly a third of its total output went to the Direct Service Industries.)[19] During World War II, the aluminum companies supplied defense producers, including Boeing in Seattle and the Kaiser Shipyards near Portland. The population of Oregon and Washington grew by over 700,000 from 1940 to 1945, boosting energy demand further.

Bonneville trumpeted its defense role with pride. "Power from the Columbia River is building the ships and plants to defend the land we love," proclaimed one poster. Another announced, "Bonneville is on the firing line!"[20] Ironically, wartime plutonium production at Hanford itself required a huge electric load – more in 1945 than all the municipal utilities, PUDs, and electric co-ops combined. The presence of a transmission line linking the two dams had been one of the features attracting Leslie Groves and Franklin T. Matthias, officer-in-charge at Hanford, to the site.[21]

By war's end, eight large generators at Grand Coulee Dam and ten more at Bonneville produced electricity. BPA's capacity grew from less than 100 megawatts in 1939 to 1350 megawatts by 1944.[22] Promoting

---

[18] See U.S. Energy Information Administration, Department of Energy, "100 Largest Electric Plants," http://www.eia.doe.gov/neic/rankings/plantsbycapacity.htm, accessed June 22, 2005.

[19] 1976 figures from Lee et al., *Electric Power and the Future*, 30, Table 4.

[20] Tollefson, *BPA and the Struggle*, 224–225.

[21] Hanford electrical load cited in Tollefson, *BPA and the Struggle*, 234. S. L. Sanger with Robert W. Mull, *Hanford and the Bomb: An Oral History of World War II* (Seattle: Living History Press, 1989), 7. For overviews of the impact of World War II on the American West, see Gerald D. Nash, *The American West Transformed: The Impact of the Second World War* (Bloomington: Indiana University Press, 1985) and Nash, *World War II and the West: Reshaping the Economy* (Lincoln: University of Nebraska Press, 1990).

[22] Norwood, *Columbia River Power*, 123, 127.

demand during the Depression and expanding supply in wartime taught the region's power planners some lessons they absorbed perhaps too well for the next generation. Hydroelectric power systems were declining cost industries with large capital requirements but great economies of scale. Taking advantage of these economies required long-term planning, optimism, and a promotional outlook. Industrialization would not only provide the level of demand that would give the area low-cost power, it would liberate the Northwest from its long history of economic dependence on natural resource extraction and colonial subservience to Eastern business and finance. For most advocates of public power, there were corollaries to these propositions. If the Northwest were to take advantage of economies of scale in power generation, large-scale investment by the federal government, coordination by Bonneville, and preferential treatment that would link small public utilities into a broad system would all be required.

Even before the war's end, Congress authorized expansion of the hydroelectric system on the Columbia and its tributaries. An order from Secretary Ickes in October 1945 assigned BPA the task of marketing the power that additional dams on the Columbia, Snake, Willamette, and Flathead Rivers would generate. Although the Republican-majority Eightieth Congress (1947–49) viewed public power warily, Bonneville escaped serious curtailment, and its budget and construction activities grew markedly in the later years of the Truman administration. By 1953, there were eleven federal hydro projects under construction in the region.

Incipient shortages spurred capacity expansion in the postwar period. Although the private utilities had predicted an era of surplus electricity, continued population growth, further expansion of aluminum production, and industrialization turned demand curves upward after a brief slump in 1945–46. With a water power system, electric output depends on rainfall and the runoff of melting snow into the river system. Dry periods spell trouble, and Bonneville had difficulty meeting peak demands in 1948; in 1951 and 1952, BPA had to interrupt power for its large industrial users, the Direct Service Industries.[23]

---

[23] Bonneville's own discussions of the 1951 and 1952 shortages suggest the military influence on power policy. The Korean War-era annual BPA Advance Programs stressed that Bonneville's need for more generating capacity was accentuated by the need for war production. Curtailments of industrial output in 1951, the Advance Program reported, had not been severe enough to affect defense production. United States Department of Interior, Bonneville Power Administration, 1952 *Advance Program for Defense* (Portland, OR: Bonneville Power Administration, 1952), 19. The addition of "for Defense" to the annual program's title is itself notable.

The Northwest's problem was exacerbated by the fact that it was and is the only region of the United States where peak demand comes in the winter, as electric heating and lighting requirements increase; the moderate climate in population centers near the Pacific Coast makes the region less saturated with air conditioning, and hence summer demand, than elsewhere. Meanwhile, however, the streamflows determining hydropower supply reach their maxima in late spring and summer. Beginning with Grand Coulee, projects on the upper Columbia and some of the feeder rivers were designed to store water in reservoirs behind dams for release when peaking power was needed, boosting output at downstream dams as well. This and other engineering techniques served to merge the separate facilities into a unified operating system. It was Bonneville's long-standing intention, articulated in the early 1960s by Administrator Charles F. Luce, to operate on a "single utility" basis.[24]

The drive for efficiency through growth dictated centralized operations at BPA, but the late 1940s and early 1950s saw a renewal of the public power–private utility strife that had been subordinated during the war. In Oregon and Idaho, restrictive legislation made it hard to carve out new PUDs. However, PUDs in Washington cut chunks of territory from the private firms' domains through negotiated sales and condemnation proceedings. In the same years, rural electric cooperatives virtually completed the task of bringing light and power to the Northwest's farms.

While the public-private power battles raged, internecine conflict among the publics also became a factor in Northwest electric politics. Especially divisive were conflicts between public utilities that owned their own power supplies and those wholly dependent on Bonneville for electricity. The long-established municipal systems in Tacoma, Washington, and Eugene, Oregon, in particular were often more interested in collaborating with the private firms than in supporting the non-generating PUDs.

"Partnership" in the region's power development became a watchword early in the Eisenhower administration. In his first State of the Union address, the new president stated, "The best natural resources program for America will not result from exclusive dependence on Federal bureaucracy. It will involve a partnership of the States and local communities, private citizens, and the Federal Government, all working together." At a March 1953 press conference, Eisenhower elaborated on this: "I just don't believe the Federal Government should be in these things except

[24] Norwood, *Columbia River Power*, 217.

where it is clearly necessary for it to come in, and then it ought to come in as a partner and not as a dictator."[25] To the private utilities, partnership offered an alternative to public power preference embedded in the Bonneville Project Act. Circumventing preference threatened what was probably public power's most persuasive selling point, its lower customer rates. Public power advocates worried the policy meant "no new starts" of federal dams, but the Eisenhower administration claimed it was not so hard and fast.

Partnership in Pacific Northwest hydropower development was to mean a variety of arrangements, none of them relying on the Roosevelt-Truman policy of federal funding, construction and ownership of Columbia River system hydro projects. If the privates built new dams, the kilowatts would be theirs. The most dramatic case came in Hells Canyon on the Snake River at the Oregon-Idaho border. The Interior Department in 1953 dropped its proposal to build a high dam at Hells Canyon to provide extensive storage for flood control and downstream power generation; instead, the Federal Power Commission accepted the Idaho Power Company's (IPC) application for a license to build five low dams without storage capacity. Facing opposition to a scheme that would not enhance downstream development, the IPC modified its plan to include three dams and a limited storage reservoir, and the administration approved this in 1955.

In Eisenhower's second term, the partnership policy sputtered to a halt. Secretary of Interior Douglas McKay, a strong backer of private power, returned to Oregon and ran unsuccessfully for the Senate. McKay's successor at the Interior Department, Fred Seaton of Nebraska, was less committed to private utility interests and more willing to support federal initiatives. Public power supporters successfully blocked a proposal to develop a huge John Day Dam on the Columbia with federal construction but private operation, calling it an unconscionable giveaway to corporate interests. Ultimately, Congress approved development of the site as part of the federal system.

The slowdown of federal dam building in the Eisenhower years while Northwest population and energy demand kept growing impelled both public and investor-owned utilities to construct their own generating capacity. Thus, the Grant County, Washington Public Utility District won

---

[25] Franklyn D. Mahar, "The Politics of Power: The Oregon Test for Partnership," *Pacific Northwest Quarterly*, 65, 1 (January 1974): 30; Norwood, *Columbia River Power*, 192.

a license in 1955 to build two large dams, Priest Rapids and Wanapum, on the Columbia below Grand Coulee. Paradoxically, although the partnership years had implied a decentralization of control of Northwest electrical energy resources, in at least two ways the region emerged from the 1950s with a more tightly knit system. First, major hydropower projects were too massive for all but the largest utilities to finance and use themselves. Thus, they had to work together with other utilities, either public or private, to raise funds and assure a market for the new supply. Public and private utilities, while often still at loggerheads, joined in cooperative ventures. For example, Douglas County (Washington) PUD and Puget Sound Power & Light agreed that the PUD would build the Wells Dam but gave Puget the first option to buy any power above Douglas's own needs. The formation of the Washington Public Power Supply System as a Washington State Joint Operating Agency in 1957 brought together PUDs in the state that were too small to build large generating plants on their own. A second force knitting the region's power system together was the need for long-distance transmission of power from non-federal projects. This pressed Bonneville into adopting a broad policy of "wheeling," transmitting power from non-federal generating facilities to its utility customers. In these ways, coordination grew during the partnership era.

The Kennedy and Johnson years brought a return to more active developmental policies in the Pacific Northwest. Walla Walla lawyer Charles F. Luce, who served as BPA Administrator from 1961 to 1966, oversaw several major enhancements to the hydroelectric power system. Yet at the same time he and other regional and national leaders pointed the Northwest toward its troubled rendezvous with nuclear energy. Luce's approach was aggressive. "In the past we've sat idly by, waiting for industry to come, and then have developed the needed power," he stated in 1962. "I want to develop power in the faith that if we have it [the power] we'll get the industry."[26]

During the Luce years, the most important expansionary venture was implementation of the Columbia River Treaty with Canada, which had been negotiated and signed at the end of the Eisenhower administration. Canada built three large storage dams on the upper Columbia and marketed its share of the added electricity to a consortium of U.S. customers.

---

[26] Quoted in Bonnie Baack Pendergrass, "Public Power, Politics and Technology in the Eisenhower and Kennedy Years: the Hanford Dual-Purpose Reactor Controversy, 1956–1962," PhD diss., University of Washington, 1974, 100–101.

The Canadian dams were to be operated as part of a unified hydropower system, internationalizing the single utility principle. The new supply gave the U.S. Pacific Northwest power that it could sell elsewhere. In September 1964, President Lyndon Johnson signed a bill authorizing a Pacific Northwest–Pacific Southwest Intertie, linking eleven states in a regional transmission grid. The Intertie has allowed seasonal exchanges between the Northwest with its winter peak demands and the Southwest, where maximum usage occurs in hot summers.[27] The Canadian agreement also allowed implementation of long-standing plans for a third powerhouse at Grand Coulee Dam and paved the way for later construction of additional generators at other dams further downstream.

In part because of the accomplishments of BPA and the growth of the federal hydropower system in the Kennedy-Johnson years, the public-private utility animosities of the New Deal and Fair Deal era were muted by the 1960s. The expanding system promised something for everybody. Moreover, the experience of cooperation in constructing new facilities during Eisenhower's administration had taught some rivals the virtues of coexistence. One private utility executive noted, "They found that we were not out to do them in and we found they were very capable and reasonable to work with." Through wheeling, implementation of the Canadian treaty and the Pacific Intertie, technical interdependence under the single utility concept also bred a degree of harmony among regional utilities.[28]

Nevertheless, long-standing conflicts remained pressing issues for many public power advocates in the Pacific Northwest during the 1960s. Ken Billington, who served for thirty years as head of the Washington Public Utility Districts Association, could admit that the investor-owned companies of the sixties and beyond were less rapacious than the absentee-owned holding companies whose tentacles had clutched and strangled the region's power supply a generation earlier. Yet in his half-century of public power advocacy, he typified those who evaluated all policies by their effect on the well-being of public power. No radical – indeed he proclaimed his support for "free competitive enterprise on which our capitalist

---

[27] Douglas Norwood, "Administrative Challenge and Response: The role of the Bonneville Power Administration in the West Coast Intertie Decision," Unpublished Bachelor of Arts thesis, Reed College, 1966. A brief account of coordination and expansion under Luce is in Tollefson, *BPA and the Struggle*, 329–345.

[28] Bruce Marvin Haston, "From Conflict Politics to Cooperative Politics: A Study of the Public-Private Power Controversy in the Pacific Northwest," PhD diss., Washington State University, 1970, 204.

economy is founded" – Billington saw public power as a democratic crusade.[29]

For some, public power in the Pacific Northwest was part of a popular social movement, combating economic colonialism and corporate domination, providing opportunity and security for ordinary American families, democratizing daily life and enhancing community self-governance.[30] These aspirations would continue to motivate and shape many of the actions of the Washington Public Power Supply System and its supporters as the System undertook its massive nuclear ventures. Another perspective, distinctly less idealistic, would see public power as part of a New Deal strategy of "state capitalist" regional development. In *The New Dealers*, Jordan Schwarz argued powerfully that the real movers and shakers of the Roosevelt era were men dedicated to using the resources of the federal government to build a public infrastructure for privately led growth in the backward regions of the South and the West. In this analysis, dynamic figures like Henry J. Kaiser, who put together the construction consortiums that built Bonneville and Grand Coulee Dams, loom large.[31] A third vantage point, one that stresses structures rather than either popular or elite agency, would interpret Pacific Northwest electrical power development as a process of system-building. The large dams, the transmission grid that spread across the region, postage stamp pricing, operation of the generating facilities under the single utility concept, coordinating and planning bodies like the Pacific Northwest Utilities Conference Committee (PNUCC), power arrangements under the Canadian treaty, and the Pacific Intertie – these were the benchmarks of evolution toward a large-scale regional power system. Paradoxically, the Northwest moved toward this complex system for producing, transmitting, and distributing energy with a fragmented organizational structure. Over a hundred consumer-owned utilities, along with a handful of private companies and a score of Direct Service Industries were increasingly tied into a regional network of dams, power lines, and coordinating bureaucracies.

[29] Ken Billington, *People, Politics and Public Power* (Seattle: Washington Public Utility Districts Association, 1988), 25.

[30] Jay Brigham, *Empowering the West: Electrical Politics before FDR* (Lawrence, KS: University Press of Kansas, 1998) interprets public power activism in the early twentieth century from this perspective. Chapter 5, "Seattle and Washington State: Focal Points of the Public Power Fight," 96–123, examines the Pacific Northwest.

[31] Jordan A. Schwarz, *The New Dealers: Power Politics in the Age of Roosevelt* (New York: Alfred A. Knopf, 1993). Chapter 14, "Henry J. Kaiser: New Deal Earth and Money Mover," 297–342, treats Kaiser as the epitome of the powerful operatives of the Roosevelt years.

A quest for technical efficiency and economic progress – along with less lofty motives of bureaucratic aggrandizement and political pork – drove the expansion of the Northwest's power system. Another force also deserves emphasis: the context of World War II and the Cold War. To repeat, it was the Second World War that had brought the demand for the power of Bonneville and Grand Coulee Dams. The economy of Washington State remained heavily militarized in the postwar era. In fiscal year 1963, it ranked eighth among the states in military prime contracts; with only 1.6 percent of the nation's population, it received 4.1 percent of the Defense Department's contract awards.[32] Henry M. ("Scoop") Jackson's congressional career moved in tandem with the state's Cold War commitments. By 1958, with approximately 60,000 Boeing Company employees in the Seattle area, opponents were already labeling him the "Senator from Boeing." For Jackson and other politicians, military concerns became rationales for expanding Northwest electricity supplies just as the Soviet menace justified federal policies ranging from interstate highway construction to college student loans.[33] In the region, growth was a watchword. A newspaper near the Hanford Reservation had put it boldly in 1950. To win the Cold War, the nation had "to develop every natural resource at its command.... The West is the last economic frontier. Its agricultural and industrial settlement is imperative."[34] And, as we shall see, the Washington Public Power Supply System staged its entry into civilian nuclear power generation by quite literally clinging to a weapons plant.

Charles Luce left Bonneville to become Undersecretary of the Interior in 1966. In less than thirty years, BPA had grown from a temporary administrative vehicle into a leading force in the Northwest's regional economy. Bonneville's service area covered 271,000 square miles, including

---

[32] U.S. Department of Defense, Washington Headquarters Services, Directorate for Information, Operations and Reports, *Department of Defense Military Prime Contract Awards by State, Fiscal Years 1951 to 1983*, 15.

[33] For Jackson's relationship to Boeing, see Richard S. Kirkendall, "The Boeing Company and the Military-Metropolitan-Industrial Complex, 1945–1953," *Pacific Northwest Quarterly*, 85 (October 1994): 137–149 and Kirkendall, "Two Senators and the Boeing Company: The Transformation of Washington's Political Culture," *Columbia: The Magazine of Northwest History*, 11 (Winter 1997–98): 38–43. Political scientist T. M. Sell has recently contended that Boeing's influence in community and state politics has been limited: T. M. Sell, *Wings of Power: Boeing and the Politics of Growth in the Pacific Northwest* (Seattle and London: University of Washington Press, 2001).

[34] Quoted in John Findlay, "Lesson 24: The Impact of the Cold War on Washington: Hanford and the Tri-Cities," http://www.washington.edu/uwired/outreach/cspn/hstaa432/lesson_24/hstaa432_24.html, accessed June 9, 2004.

all of Washington, Oregon, and Idaho, Montana west of the Continental Divide, and small portions of Wyoming, Utah, Nevada, and California. Twenty-one completed projects and another dozen under construction or authorized comprised the Federal Columbia River Power System. The capacity of 6,678 megawatts amounted to 44 percent of the region's total power supply. In a region with less than 3 percent of the nation's population, the system contained 15 percent of the installed hydro capacity. If projects under construction, authorized or licensed were completed, the federal system would grow to 20,708 megawatts, or 65 percent of expected generating capacity by 1976.[35] This was a major enterprise: Bonneville revenues in its first fiscal year had reached only $49,835; in fiscal year 1967, federal system income was nearly $113 million.[36]

Bonneville had provided cheap electricity; a 3 percent rise in 1965 had been the only rate increase in the agency's history. Although the national average residential cost per kilowatt-hour was 2.20 cents in 1966, most public utilities in the Pacific Northwest kept residential rates below a penny. Cheap electricity encouraged liberal use. The region's residences in 1965 consumed an average of 11,200 kilowatt-hours yearly; the United States mean was only 4,900. Northwest electrical utilities had promoted the all-electric home, using electricity for heating and cooking as well as illumination. Commercial and industrial loads had grown rapidly as well. About 30 percent of the nation's aluminum reduction capacity had located in the Northwest, drawn by low-cost energy.[37]

The 1960s, then, saw an impressive culmination to an era of hydropower development. Those who had participated in building the system took justifiable pride in their accomplishments. They comprehended their mission in large, even grandiose, terms. In 1967, Gus Norwood, the Executive Secretary of the Northwest Public Power Association, speaking to an International Conference on Water for Peace, repeated Pope Paul VI's statement, "Development is the new name for peace." He reached as well for the maxim of architect Daniel Hudson Burnham: "Make no little plans."[38]

Those who preached the blessings of development also warned of the dangers of failing to move forward. The growing population of the Northwest, new residential uses for electric power, and increased business

[35] United States Department of the Interior, Bonneville Power Administration, *Advance Program 1966–1976* (Portland, OR: Bonneville Power Administration, 1966), 7, 11.
[36] *Advance Program: 1967–1987*, 2.
[37] Lee et al., *Electric Power and the Future*, 38.
[38] Gus Norwood, "Public Objectives in Water Resources Development," *Northwest Public Power Bulletin*, 21, 5 (May 1967): 7, 9, 11.

activity would all make demand grow rapidly. Each year, they predicted, the region would need 1,000 more megawatts of power, enough to supply another city the size of Seattle and about twice the existing capacity of Bonneville Dam. Yet, they noted, the river system would soon approach its maximum generating capacity. With the completion of John Day Dam in July 1968, there was only one sizeable free-flowing stretch of the Columbia, below the Priest Rapids Dam, through the Hanford Reservation and past the Tri-Cities. New dams on tributary rivers could provide additional hydroelectric power, but not enough to meet predicted needs. Deficits would appear by the early 1970s; the hydroelectric system, the region's glory, would no longer be able to keep the lights bright and the machines running.

There was a solution.

Power and Development: The National Context

Producing and distributing electricity combines some of the characteristics of heavy manufacturing with many features typical of service industries. Electric generating utilities are extremely capital-intensive, with high fixed costs and low variable costs, similar to the pattern in many large manufacturing enterprises. The main variable cost in fossil fuel–powered systems is, of course, fuel; in hydropower generation, falling water substitutes for fuel in providing the energy for conversion to electricity. For a nuclear plant, operations and maintenance costs will be greater than the cost of fuel. Because new plants involve substantial amounts of capital, they normally require external financing. As with most capital-intensive undertakings, investment in electric utilities is often "lumpy." A new generating facility will often provide more electricity than immediate needs require. Moreover, these investments are often technically complex and require substantial planning and construction time. Thus, utilities must plan and forecast; they cannot simply respond to immediate market developments. In their important 1980 study of *Electric Power and the Future of the Pacific Northwest*, Kai N. Lee, Donna Lee Klemka, and Marion E. Marts invoked John Kenneth Galbraith's portrait of the "planning system" in *The New Industrial State* to account for the anatomy and behavior of regional electric power.[39] Although Lee and his co-authors

---

[39] Lee et al., *Electric Power and the Future*, 89–96; John Kenneth Galbraith, *The New Industrial State* (2d ed., revised; Boston: Houghton Mifflin, 1971). In recent years, technological changes, rising fossil fuel costs, and partial deregulation have altered these characteristics of electrical utilities somewhat, but the basic pattern still holds.

stressed the distinctiveness of the Northwest's regional situation, electric power industries in industrialized societies have for decades shown these characteristics.[40]

Galbraith's model derives from manufacturing, but electric utilities differ from manufacturing firms in some basic respects. Perhaps most significantly, utilities must deliver their intangible output upon demand. (There are exceptions, such as the interruptible share of the loads of Northwest Direct Service Industries, discussed below, but these are not the norm.) Kilowatts cannot be inventoried or warehoused; they are produced and delivered virtually simultaneously with their consumption. This means that utilities need to have the capacity to supply the maximum amount demanded at any one moment. Utilities must pay close attention to their load factor, the ratio of the average amount demanded in a given period of time to the peak demand in that period. A low load factor means a large share of invested capital will lie idle. Because electric utilities are capital-intensive, it is costly to run below full capacity. Load factor is a key indicator of efficiency.*

To achieve a high load factor, utilities in the United States and Europe since at least World War I have sought interconnections. If one group of customers (say, residential ratepayers) reaches its peak power demand in early evenings while other users need power during the workday (commercial and office clients), there is strong incentive to supply both groups from the same sources to keep those facilities more steadily in use. A utility serving suburban householders will also want commercial and industrial customers. Alternatively, different utilities can serve each load efficiently if they arrange to share common sources of power. Seasonal variations in electricity loads can also be smoothed out with interconnections. Where air conditioning is in widespread use, summer peaks will substantially exceed winter demand; in a cooler region like the Pacific Northwest with widespread electrical home heating, maximum demand will come in the winter. Interconnecting the two regions can raise the load factor for each. These considerations are crucial for central-station power systems. In the words of Thomas P. Hughes, "During the twentieth century expansion for

[40] Thomas P. Hughes, *Networks of Power* (Baltimore: The Johns Hopkins University Press, 1983) demonstrates how sociocultural, as well as technical, forces shaped the development of large-scale, complex electrical utility systems in the United States, the United Kingdom and Germany.
* Yet a very high load factor indicates potential strains on utility capacity. Facilities will inevitably need maintenance and repair, often requiring curtailed output. Without a cushion between average load and peak demand, service interruptions are a danger.

diversity and management for a high load factor have been prime causes for growth in the electric utility industry.... The load factor is, probably, the major explanation for the growth of capital-intensive technological systems in capitalistic, interest-calculating societies."[41]

Reliability is another rationale for interconnection. Any generating facility is subject to shutdowns; natural catastrophe, technical or organizational failure, or simply the need for maintenance can cause a plant to go off line. When this occurs, a link to other generating plants will usually help keep the power flowing.

The cost structures of different power sources greatly affect power system planning. Hydroelectric power is exceptionally capital intensive, even in comparison to other generating systems. Thus, a hydro generating facility costs almost as much when it is out of service as when it is producing at full capacity. This suggests using large-scale hydropower to meet a system's base load, the power demand it can expect throughout normal periods of operation. Smaller, less capital-intensive facilities can be operated to meet peak demands. The marginal cost of electricity from these generators (in recent years, most often gas-fired combustion turbines) is higher, but the fixed cost incurred while they are inactive is relatively low. With this mixture of resources, a utility can maximize its "firm" capacity, energy that customers can count on without danger of interruptions or curtailed service.

To make hydropower and nuclear reactors complementary sources of supply, some modifications of this fundamental planning strategy were needed. First, hydroelectric systems can be operated to meet peak loads. Water stored behind dams during periods of low usage can be released to provide additional power when needed, both at the storage facility itself and downstream at other generating dams. It may even pay to pump spilled water back up to a storage reservoir to release for electricity when demand is high. Second, although the high capital costs of hydroelectric plants speak for operating them as base load facilities, it is relatively easy to vary the amount of energy they generate by controlling water flow. On the other hand, nuclear reactors cannot be rapidly started up or shut down, and the process entails operating costs and stresses on the fuel rods. Therefore, if a system mixes hydroelectric power with large nuclear

---

[41] Thomas P. Hughes, "The Evolution of Large Technological Systems," 51–82 in Wiebe E. Bijker, Thomas P. Hughes and Trevor J. Pinch, eds., *The Social Construction of Technological Systems* (Cambridge, MA: MIT Press, 1987), 72.

reactors that cannot readily be turned on and off, it could make technical and economic sense to use the nuclear plants for base load generation and look to hydropower to meet peak demands.

By the late 1960s, American utilities had over three-quarters of a century of experience in generating electricity, transmitting it from generating sources, and distributing it to customers. Installed generating capacity in the United States had grown from about 1.2 million kilowatts in 1902 to 291 million kilowatts in 1968. Utility plant assets in the latter year reached nearly $65 billion, about $323 per capita, nearly half the net value of capital in the entire manufacturing sector of the economy. Electrification of urban areas had come quite rapidly early in the century; nearly 85 percent of the nation's non-farm residences had current by 1930. In that year, however, only a tenth of American farms received electric service. By 1950, over three-quarters had, and rural electrification was virtually complete by the 1960s. Thus, tens of millions of adults in 1968 who had grown up on farms could remember the day the power came to their homes. Consumption per household had nearly doubled in the previous decade, and the price of electricity had continued to fall. A kilowatt-hour cost residential users an average of 6.03 cents in 1930, 3.01 cents in 1948 and only 2.12 cents in 1968.[42]

These accomplishments resulted from steady evolution toward large-scale social and technical systems of electric power supply. Complex organizations, often vertically integrated from generation through transmission to retail distribution, produced electricity. They invested heavily in capital-intensive facilities with high fixed but low variable costs per unit of output. Since the early twentieth century, utility economics had been predicated on the assumption that this was a declining cost industry, that marginal cost (the cost of generating an additional unit of electricity) would go down as facilities got larger and output went up. This in fact had been the basic justification for government regulation of privately owned utilities. They were thought to be "natural monopolies," and it was the task of objective regulators serving the public interest to ensure that the efficiencies of monopolistic large-scale enterprise were passed on to the rate-paying consumers. Advocates of public power contended that regulation alone would not assure the fair distribution of those benefits,

[42] United States Bureau of the Census, *Statistical History of the United States, from Colonial Times to the Present* [originally published as *Historical Statistics of the United States, Colonial Times to 1970*] (New York: Basic Books, 1976), 824, 829, 258, 827. Figures on capacity refer to utilities, excluding industrial plants producing their own electricity.

but few if any – public or private – doubted the blessings of expanded size and scope.[43]

Electrical engineers, who had often moved into managerial positions in electric utilities, agreed there were great advantages to large-scale generation facilities. As Richard Hirsh has shown, the utility industry's conviction that ever-larger plants would mean declining costs per kilowatt of capacity held true into the 1960s. One rule of thumb held that a 0.6 percent increase in capital outlay would increase capacity by 1.0 percent. Another measure of technical progress was thermal efficiency, the proportion of fuel energy transformed into electrical energy. This had grown from about 2.5 percent in the first central generating stations to 32.9 percent by 1965.[44]

In the 1960s, the utility industry's faith in economies of scale was strongest where it had been least tested, in nuclear power plants. Their experience derived overwhelmingly from coal-fired generation, but executives and engineers were almost certain that ever-larger nuclear plants would push nuclear electricity down to competitive cost levels. They consistently ordered nuclear plants larger than any that had yet been built. In the Pacific Northwest, planners posited that the cost of electricity from a nuclear plant of 200 megawatts would be 40 percent higher per megawatt than from a 1,000 megawatt reactor. However, in 1968, when these plans were laid, there were no operating nuclear plants in the world larger than 600 megawatts.[45]

After nearly two decades of hesitations and false starts, commercial nuclear power in the mid-1960s seemed to be taking off. In December 1963, General Electric Company and Jersey Central Power and Light had signed a pioneering contract for a 515 megawatt nuclear plant at Oyster Creek, New Jersey. Oyster Creek was to be a "turnkey" plant. GE took responsibility for building the entire facility at a set price by a given date

---

[43] Thomas K. McCraw, *Prophets of Regulation* (Cambridge, MA, and London: Harvard University Press, 1984), 231–243 contains an incisive discussion of the political and intellectual history of the concept of regulation in the public interest. Richard F. Hirsh, *Technology and Transformation in the American Electric Utility Industry* (Cambridge, UK and New York: Cambridge University Press, 1989) provides an excellent account of the period of declining costs and the forces that were to reverse the trend.

[44] Hirsh, *Technology and Transformation*, 40–46, 89–90.

[45] BPA, *Hydro-Thermal Power Program*, 13–14. U.S. Federal Power Commission, *Steam-Electric Plant Construction Cost and Annual Production Expenses*, Twenty-First Annual Supplement – 1968 (Washington, D.C.: U.S. Federal Power Commission, 1969), xiii, 151. (The Hanford Generating Project, completed in 1966, was rated at 800 megawatts, but the reactor was primarily for producing plutonium, not electricity.)

and turning it over to the utility ready to operate. This was the first nuclear plant ordered without direct subsidy by the Atomic Energy Commission. General Electric and Jersey Central startled the utility industry with the proposition that within five years the plant would produce cheaper electricity than any conventional fossil fuel project. Westinghouse Electric, faced with its giant competitor's new marketing strategy, soon followed suit. Eventually, the manufacturers sold thirteen plants on a turnkey basis. By assuming the risks inherent in large-scale construction and engineering projects, General Electric and Westinghouse coaxed a civilian nuclear power market into being after nearly two decades of electrical utility skepticism. As one GE executive explained, the turnkeys were needed if the companies were to get returns in the civilian market for their extensive investments in nuclear submarine reactors for the U.S. Navy:

We had a problem like a lump of butter sitting in the sun. If we couldn't get orders out of the utility industry, with every tick of the clock it became progressively more likely that some competing technology . . . would supersede the economic viability of our own. . . . [I]f we didn't force the utility industry to put those stations on line, we'd end up with nothing.[46]

By 1965, utilities were willing to invest in nuclear projects without fixed price guarantees. In 1966–67, they ordered forty-nine nuclear plants with nearly 40,000 megawatts' capacity. Philip Sporn, former President of the huge American Electric Power System, who remained a doubter, called this the "Great Bandwagon Market." The enthusiasm of nuclear supporters justified the label.[47] Physicist Alvin Weinberg, Director of the Oak Ridge National Laboratory, foresaw a "nuclear energy revolution . . . based upon the permanent and ubiquitous availability of cheap power."[48]

The turnkey plants and the other reactors ordered in the mid-1960s turned out to be costly ventures. According to Irvin Bupp and Jean-Claude Derian, light water reactors ordered through the end of the decade cost twice as much in constant dollars as their initial estimates.[49] The turnkeys in effect had been loss leaders for GE and Westinghouse, who accepted

---

[46] Hertsgaard, *Nuclear Inc.*, 42–43.

[47] Irvin C. Bupp and Jean-Claude Derian, *Light Water: How the Nuclear Dream Dissolved* (New York: Basic Books, 1978), 42–50; Hertsgaard, *Nuclear Inc.*, 40–49. Sporn's doubts can be found throughout his writings and talks in the 1960s. See, in particular, Philip Sporn, *Technology, Engineering, and Economics* (Cambridge, Mass.: MIT Press, 1969), 49–50.

[48] Bupp and Derian, *Light Water*, 50.

[49] Ibid., 79.

this as the price of establishing themselves in the market ahead of domestic and international competitors.

The untested dogma of economies of scale in nuclear plants was one cause of the cost overruns. Utilities consistently ordered plants to be far larger than any operating reactors. Between 1963 and 1967, the largest reactor in operation had a capacity of 200 megawatts. Yet the smallest new plant ordered in those years was over 400 megawatts.[50] More recently, in analyzing the stagnation of the American nuclear power industry, commentators have emphasized the failure to take advantage of past experience. Plant sizes leapfrogged each other as the nuclear power industry put its trust in scale economies rather than standardized design and the value of learning from past practice.[51] The utility industry as a whole after World War II ordered ever-larger plants with technologies that had not yet been widely tested in use. The manufacturers practiced "design by extrapolation" to provide these latest-model facilities.

Until late in the 1960s, the approach had paid off, although there had been warning signals. Several of the most modern and largest facilities had developed unexpected, serious operating problems. By the 1970s it was becoming evident that the newest and biggest plants were not always the best. As one utility executive put it, "We hoped the new machines would run just like the old ones we're familiar with. They sure as hell don't." Technological barriers to increased thermal efficiency and economies of scale bespoke what Richard Hirsh called "technological stasis." He emphasized that this syndrome of a mature utility industry is "not just a hardware problem. It is a *systems* phenomenon that comprehends technical and *social* components."[52] The choices and motives of different interest groups (manufacturers, utilities, regulators, and others) interacted with the purely technical factors to bring stasis on. Yet as stasis loomed, utilities in the late 1960s went on a plant-buying binge, ordering 129 new turbine generators in 1966 alone. This represented more than twice the

---

[50] See chart in Bupp and Derian, *Light Water*, 73.

[51] For commentaries on the structural problems of the civilian nuclear power industry in the years after the nuclear bandwagon ground to a halt, see Peter Stoler, *Decline and Fail: The Ailing Nuclear Power Industry* (New York: Dodd, Mead, 1985); Morone and Woodhouse, *Demise of Nuclear Energy*; Campbell, *Collapse of an Industry*; James Cook, "Nuclear Follies," *Forbes*, 135, 3 (February 11, 1985): 82–100; Robin Cowan, "Nuclear Power Reactors: A Study in Technological Lock-in," *Journal of Economic History*, 50, 3 (September, 1990): 541–567.

[52] J. Samuel Walker, *Containing the Atom: Nuclear Regulation in a Changing Environment, 1963–1971* (Berkeley and Los Angeles: University of California Press, 1992), 34; Hirsh, *Technology and Transformation*, 189.

capacity ordered in any previous year.[53] In 1967 and 1968, nuclear plants represented nearly half of the new capacity ordered.[54]

A few in the electrical power industry recognized the impending problems. One of its leading executives, Philip Sporn, kept up a steady flow of questioning, prodding evaluations of industry practices and looked skeptically at rosy forecasts of nuclear power's efficiency. (That his firm, American Electric Power, served the coal-mining region of the Ohio River Valley may not have been coincidental.) In 1967, he expected that nuclear capacity would reach about 115,000 megawatts by 1980, far less than the 180,000 megawatts that the Atomic Energy Commission was predicting. (In fact, actual nuclear capacity in 1980 was only about 51,000 megawatts.[55])

In retrospect, the electric utility industry in the late 1960s gives the impression of scanning the wrong horizon for the problems it would soon encounter. Sympathetic summaries of industry difficulties appearing in *Fortune* and *Business Week* in 1969 noted rapidly growing demand and the need for capacity expansion, shrinking reserves, construction delays, and cost overruns due to material and labor shortages, some financing difficulties, and problems dealing with pesky conservationists and regulators.[56] Few utility leaders seemed prepared for the business environment emerging in the 1970s: technological stasis, grave disappointments in nuclear generation, an energy crisis, a drastic decline in demand growth, and far broader challenges to a supply-oriented industry than its executives had imagined.

The Northwest Plans an Energy Future: Toward a Nuclear Northwest

The hydroelectric legacy, the intense controversies over public power, the role of Bonneville, and the fact that the region paid the nation's lowest rates and used the most electricity per capita made the Northwest's electrical energy situation distinctive in the late 1960s. However, regional power planners and decision-makers echoed many of their industry counterparts nationally. Anticipated demand increases, extensive planning for

---

[53] Ibid., 94.   [54] Walker, *Containing the Atom*, 34.

[55] Sporn, *Technology, Engineering, and Economics*, 58–59; U.S. Bureau of the Census, *Statistical Abstract of the United States: 1988* (108th edition.) (Washington, D.C.: GPO, 1987), 547, Table 930.

[56] Jeremy Main, "A Peak Load of Trouble for the Utilities," *Fortune*, 80, 6 (November 1969): 116–119, 194–205; "Why utilities can't meet demand," *Business Week*, no. 2000 (November 29, 1969): 48–62.

long-term capacity growth, strong belief in economies of scale through both large plants and system expansion, and enthusiasm for nuclear power were hallmarks of the Northwest's energy strategies.

Technically, the Northwest could have expanded hydropower output in the late 1960s. Admittedly, there were few if any sites to build more of the major dam projects that dotted the Columbia and the Snake. Yet, in a 1970 speech, Owen W. Hurd, the first managing director of the Washington Public Power Supply System and an enthusiast for nuclear development, pointed out, "The all hydro era came to an end with less than fifty percent of the region's hydroelectric potential developed.... It is apparent that in retrospect, failure to make timely use of more of this renewable, non-polluting and multi-purpose low cost energy source will prove to be a tragic mistake...."[57] Hurd's assessment of hydropower neglected potentially severe environmental impacts of its further development, but in retrospect the leap into nuclear development and its downfall in the early 1980s gives poignancy to his musings.

In the late 1960s, one potential location for a major hydro project was the High Mountain Sheep site on the Middle Snake, below the Idaho Power Company's Hells Canyon Dam. Plans for developing High Mountain Sheep dated from the 1950s but had become entangled in conflicts between public and private utilities. The Supply System itself had become a protagonist in the struggle in 1960, when it filed an application for a license with the Federal Power Commission to build a power dam at a nearby site, Nez Perce. By the time the complex legal case reached the U.S. Supreme Court in 1967, those involved expected a decision to authorize a major dam on the Middle Snake, built either by the Supply System or the Pacific Northwest Power Company (a construction consortium of regional private utilities) or even the Federal Interior Department's Bureau of Reclamation. However, in a 6–2 decision, with the Northwest's own Justice William O. Douglas writing for the majority, the Supreme Court sent the case back for rehearing to the FPC with a strong suggestion that there be no dam at all.

Douglas's opinion forcefully introduced the environmentalist concerns that characterized the jurist's life and thought. A determination of whether a license is in the public interest, he wrote, "can be made only after an exploration of all issues relevant to the 'public interest,' including future

---

[57] Owen W. Hurd, "Breeder Reactor Program Report," Presented to Northwest Public Power Association Engineering and Operation Annual Conference, Idaho Falls, June 18, 1970, p. 1.

power demand and supply, alternate sources of power, the public interest in preserving reaches of wild rivers and wilderness areas, the preservation of anadromous fish for commercial and recreational purposes, and the protection of wildlife."[58]

Douglas's warning jarred public and private power alike into a partnership proposal. In the fall of 1967, WPPSS agreed with Pacific Northwest Power to petition the Federal Power Commission to license a joint dam project on the Middle Snake. These uneasy allies continued to press their case, but the political currents were too strong. This plan ran afoul first of Interior Department interest in building a federal dam at High Mountain Sheep and then of environmental concerns that eventually blocked the project. In 1975 Congress passed legislation creating the Hells Canyon National Recreation Area and effectively blocking dam construction. Another hydropower project, the Ben Franklin Dam, was planned for the mid-Columbia. It would have dammed up the last free-flowing stretch of the river, ironically the Hanford Reach section passing through the Hanford Reservation. Environmentalist opposition eventually killed the proposal in Congress in 1971.

Both the High Mountain Sheep site and the Hanford Reach still sit undeveloped. Environmentalists, sport fishing enthusiasts and other recreationalists, and Native American tribes had become players in Pacific Northwest energy politics by the 1970s. By their reckoning, large dams were anything but the unalloyed blessing that their advocates had claimed. The "great, rushing torrent of the Columbia" had been converted into "a series of slightly muddy lakes," commented Oren Bullard in an angry tract, *Crisis on the Columbia*, published in 1968.[59]

With little hope for new base load hydroelectric generation and with strong demand growth taken as a given, Northwest power interests looked to a new era of thermal generation. From 1951 on, in fact, the Bonneville Power Authority had investigated the possibility of steam generation in the region. Its Advance Program for Defense (note again the military justification for development) that year suggested that the federal government

[58] Udall v. Federal Power Commission, 387 U.S. 428 (1967), 450. William Ashworth, *Hells Canyon: The Deepest Gorge on Earth* (New York: Hawthorn Books, 1977) provides a thorough account of the legal and political controversies over damming the canyon. Chapter nine, "High Mountain Sheep: The Second Nez Perce War," focuses on the Supply System's role.

[59] Oral Bullard, *Crisis on the Columbia* (Portland, OR: Touchstone Press, 1968), 16. Bullard also found nuclear power flawed; waste heat discharged into the river would damage fish habitats.

build three 100-megawatt plants in Oregon and Washington to operate continuously during high demand winter months and when needed at other times to meet peak demands. Scoop Jackson introduced a bill in the House of Representatives to authorize $60 million for eight thermal plants to become part of Bonneville's system.[60] These plans were shunted aside, but the expectation that thermal plants would eventually complement the hydropower system remained. As early as 1955, Bonneville's annual planning document contained a discussion of potential thermal generation projects and noted that, "At the invitation of the Atomic Energy Commission, BPA has participated in research and basic engineering investigations of nuclear reactors." At that point, the report noted, atomic energy would not compete in cost with hydropower for many years.[61]

By 1958, Bonneville's emphasis had shifted. BPA's program that year noted that initially steam generation would come from fossil fuels but, "The advent of nuclear-electric energy will be hastened by the comparatively high cost of fossil fuels in the area."[62] Soon Bonneville and other energy development advocates in the region were working to make this prophecy a reality. The Washington Public Power Supply System, through a mixture of planned growth and historical accident, was to become public power's vehicle to realize its nuclear visions.

Nuclear power in the Northwest received a boost when President Kennedy, upon the advice of Scoop Jackson, named Charles Luce as BPA Administrator in January 1961. As we have seen, Luce's term at Bonneville, 1961–66, reinvigorated the agency after the hesitations of the Eisenhower years. Meanwhile, Luce and Jackson, backed by nuclear energy enthusiasts in the nation's capital such as Glenn Seaborg at the Atomic Energy Commission and Senator Clinton Anderson, chairman of the Joint Committee on Atomic Energy, pressed for a unique method of introducing nuclear electric generation at the Hanford Reservation. This was the Hanford Generating Project. It would use the heat from the so-called N-Reactor, a military plutonium-producing reactor, to generate

---

[60] U.S. Department of the Interior, Bonneville Power Administration, *1951 Advance Program for Defense* (Portland, OR: Bonneville Power Administration, 1951), 39. Marquis Childs, *The Farmer Takes a Hand: The Electric Power Revolution in Rural America* (Garden City, NY: Doubleday, 1952), 229.

[61] U.S. Department of the Interior, Bonneville Power Administration, *1955 Advance Program* (Portland, OR: Bonneville Power Administration, 1955), 30.

[62] U.S. Department of the Interior, Bonneville Power Administration, *Advance Program, U.S. Columbia River Power System, 1957.8–1968.9* (Portland, OR: Bonneville Power Administration, 1958), 21.

electricity. In the political compromise – discussed in chapter 2 – which won approval for the scheme, the Washington Public Power Supply System financed, built, and operated the turbine generators but half the power was reserved for private utilities.[63]

The Hanford Generating Project brought public and private power interests in the Pacific Northwest closer together. Like the mid-Columbia dams that some of Washington's PUDs built, the Project brought governmental investment supplying energy to private as well as public utilities. The public utilities, especially those lacking their own generating facilities, continued to rely on the preference principle for power from the federal hydro system, but they were not averse to cooperating with the investor-owned companies on projects the federal government would not undertake. Following upon the Canadian treaty and the Southwest Intertie, the Hanford Generating Project exemplified, in the phrase of Bruce Haston, a transition from "conflict politics to cooperative politics" in Northwest energy policy.[64]

This concord, never complete, grew out of common organizational problems and goals Northwest private and public utilities shared. But it also stemmed from national political trends. As Louis Galambos and Joseph Pratt point out, single-industry federal regulatory agencies tended to form symbiotic relations with the business sectors they regulated, leading in the postwar era to "stable patterns of growth."[65] Liberals of the Kennedy-Johnson era shied away from anti-business appeals and muted objections to monopoly and market power. Indeed, liberal theorists like John Kenneth Galbraith considered antitrust to be an anachronism in an economy whose balance depended on concentrated power groupings countervailing each other.[66] At the same time, the Kennedy and Johnson administrations employed a rhetoric of efficiency and expertise in both domestic and international affairs. For John F. Kennedy, economic policy was a matter of finding the most effective means to agreed-upon ends and "fine-tuning" fiscal and monetary policy to assure growth and price

---

[63] Pendergrass, "Public Power, Politics and Technology" provides a detailed account of the politics of the dual-purpose reactor.

[64] Haston, "From Conflict Politics," 288.

[65] Louis Galambos and Joseph Pratt, *The Rise of the Corporate Commonwealth* (New York: Basic Books, 1988), 152.

[66] John Kenneth Galbraith, *American Capitalism: The Concept of Countervailing Power* (Boston: Houghton Mifflin, 1952); see also Richard Hofstadter, "What Happened to the Antitrust Movement?" 113–151 in Earl F. Cheit, ed., *The Business Establishment*, first pub. 1954 (paperback ed.; New York: John Wiley & Sons, 1957).

stability. (Of the two objectives, it was growth that inspired enthusiastic commitment. Thus, signs at the U.S. Department of Commerce asked employees, "What have you done for Growth today?"[67]) For Secretary of Defense Robert S. McNamara, remodeling the Defense Department took the same kind of technical and organizational skills he had applied at the Ford Motor Company. The politics of consensus that Lyndon Johnson championed collapsed as the Vietnam War escalated, but it represented a moment of hope that economic growth and adept political leadership could mute social conflict.

For energy leaders in the Pacific Northwest in the 1960s, harmony would be the basis for growth; growth, in turn, was a foundation for the rapprochement among public and private utility interests. Despite decades of disputes, the Northwest had seen increased technical and operational integration of utility systems, the evolution of large-scale, hierarchical organizations, and professionalized long-term planning. Systems looked like solutions. Nuclear energy could be the technology of the electrical supply system of the future. The Washington Public Power Supply System could be the organization at the center of that system.

---

[67] James Tobin, *The New Economics One Decade Older*, 13, cited in Robert M. Collins, *More: The Politics of Economic Growth in Postwar America* (New York: Oxford University Press, 2000), 52. Chapters two and three of Collins provide an excellent account of growth politics in the Kennedy-Johnson era.

2

WPPSS Steps Forward

The story of the Washington Public Power Supply System is a tale of modest beginnings and grandiose schemes. The organization that took on the task of building five large nuclear power plants had more than a decade's history as a somewhat marginal actor on the Pacific Northwest utility stage. Small though it was at first, the Supply System was ambitious, and its aspirations jibed with the region's perceived energy needs. During the years 1968–73, the Supply System undertook its first three plants under a complex set of arrangements linking the projects to the BPA's Hydro-Thermal Power Program.

The Supply System's institutional origins reflect the skirmishes between public and private power interests in the state of Washington during the 1950s. With the Eisenhower administration's aversion to new Federal hydropower projects, Northwest utilities jockeyed for control of dam sites and for assured access to the kilowatts that new projects could generate. To coordinate power development in Washington, the state legislature established a Washington State Power Commission in 1953. The five-person Commission included three representatives from public utilities and two from the state's investor-owned firms, Washington Water Power and Puget Sound Power & Light. The legislation also permitted two or more municipalities and public utility districts to form new power agencies. These joint agencies could, with the Commission's approval, build and operate generating and transmission facilities.

The law represented a compromise. Public utility districts wanted the right to form joint operating agencies, but they feared a state commission that would itself build generating facilities. As Ken Billington of the Washington PUD Association put it, "You never put your power supply (or the permission to establish a supply arm) in an outside body."[1] This was a recurrent motif in public power circles for decades. Negotiations in Governor Arthur B. Langlie's office produced the arrangement that accepted a commission but gave it authority to approve joint agency applications. The bill passed the State Senate 42–0. Unanimous support led one senator to dub this the "brotherly love" bill.[2]

Public power interests soon came to regret the deal they had struck. During Governor Langlie's Republican administration, the State Commission obstructed PUD development plans. In January 1954, 21 public utilities applied to form a joint operating agency. The Commission maintained that the utilities would have to outline plans to construct a specific project before it could approve the application, something the utilities were unprepared to do.

Worse, from the public power standpoint, the Grant County Public Utility District that summer came up against the State Power Commission over its plan to build a large dam at Priest Rapids on the Columbia. Although the Army Corps of Engineers had initially planned to build on the site, in 1952 Grant PUD had stepped in and pressed for authorization to undertake the project. On July 13, 1954, Congress passed a measure removing Priest Rapids as a federal project but leaving wide open the question of which regional or local institution might be licensed to build there. The next day, Grant PUD applied to the Federal Power Commission for a preliminary permit for construction planning. Almost immediately, however, the Washington State Power Commission intervened, filing its own FPC application for study of Priest Rapids. The PUDs felt trapped. By rejecting their joint operating agency license, the Power Commission had blocked the possibility of a collaborative project. By contesting the Grant County application, the Commission had challenged an individual utility's proposal. Grant sued, charging that the State Power Commission could develop projects only if no Public Utility District chose to apply. In February 1955, Washington's Supreme Court ruled in favor of the PUD.

Meanwhile in 1955, the Washington Public Utility Districts Association under Ken Billington pressed legislation to deny the State Power Commission the right to disapprove of joint operating agency requests.

---

[1] Dan Seligman, "The Washington Public Power Supply System," Draft report prepared for Dr. Kai N. Lee, July 1978, p. 23, in Neil O. Strand Papers, Box A1.4 (B), File Folder S095, Legal Office, Washington Public Power Supply System, Richland, Washington. (Hereafter cited as Strand Papers.)

[2] Dan Seligman, "The Washington Public Power Supply System: A Report Prepared for Dr. Kai N. Lee," University of Washington Institute for Environmental Studies, Seattle, Revised December 1978, p. 21.

This passed, according to Dan Seligman, because Billington threatened that public power supporters would hold up the Governor's proposed $100,000 budget for the Commission without the concession. By this point, Grant County Public Utility District was far enough along in planning for Priest Rapids that the other PUDs decided to let that district go it alone. However, the Washington Public Utility Districts Association had other reasons to proceed with forming a joint operating agency. There were other dam sites to fill on the Columbia and its tributaries. Public power was also already interested in thermal generation. In 1955, the Atomic Energy Commission (AEC) had announced its Power Demonstration Reactor Program, an effort to move nuclear energy toward commercial application. Hydropower gave the Northwest a regional competitive edge, but nuclear energy might, if costs continued to fall, complement the water power system. Only a year before AEC Chairman Lewis Strauss had predicted that atomic energy would make electricity "too cheap to meter." In 1956, the distinguished mathematician and economist John Von Neumann predicted that with fission and more advanced forms of nuclear energy, "a few decades hence energy may be free – just like the unmetered air."[3]

National power politics gave another reason for interest in a joint operating agency. President Eisenhower's attitude toward federal power development had never been enthusiastic. Asked for an example of "creeping socialism" at a 1953 news conference, he had pointed to the expansion plans of the Tennessee Valley Authority. He had, in fact, told his cabinet, "I'd like to see us sell the whole thing [TVA], but I suppose we can't go that far."[4] In 1956, the President had called for deep cuts in TVA's budget, prompting Democratic complaints that he wanted to dismantle all federal power agencies. Northwest public utilities feared Bonneville's transmission network would be put up for sale upon Eisenhower's re-election. If so, a regional joint operating agency of public utilities should be "standing ready and waiting to purchase all of the Bonneville transmission system within the State of Washington."[5]

---

[3] Strauss's prophecy in Gerard H. Clarfield and William M. Wiecek, *Nuclear America: Military and Civilian Nuclear Power in the United States, 1940–1980* (New York: Harper and Row, 1984), 277; Von Neumann, "Can We Survive Technology?" in David Sarnoff et al., *The Fabulous Future: America in 1980* (New York: E. P. Dutton, 1956), 37.

[4] Allen Drury, "Eisenhower's Four Years," *New York Times*, July 27, 1956; Emmet John Hughes, *The Ordeal of Power: A Political Memoir of the Eisenhower Years* (New York: Atheneum, 1963), 152.

[5] Billington, *People, Politics & Public Power*, 125.

Therefore, on June 29, 1956, the Washington PUD Association Board authorized sponsorship of a joint operating agency, which it named the Washington Public Power Supply System. By the August 1 deadline the Board had imposed, seventeen of the nineteen electric utilities in the association had resolved to join the new agency. Complying with state law, the applicants advertised in newspapers in each of Washington's thirty-nine counties their intention to construct a thermal plant and several small hydroelectric projects, although, according to Billington, the founders wanted above all to keep Bonneville's transmission network in public hands. That November voters elected Albert Rosellini, a Seattle-area Democrat favorable to public power, as Governor. Earl Coe, Rosellini's appointee as Director of Conservation and Development, gave his approval to the formation of the Supply System as a Joint Operating Agency as his first official act, on January 31, 1957.

A photo of Coe signing the authorization suggests something of the mood and the political forces at the inception of the Washington Public Power Supply System. Half a dozen men stand behind him, all middle-aged and in suits. Their faces indicate determination and seriousness of purpose. They included A. Lars Nelson, Master of the Washington State Grange; Clyde Riddell, President of the Washington PUD Association; two commissioners from rural county PUDs and a manager of a third district; and Jack Cluck, attorney with the Seattle firm of Houghton Cluck Coughlin and Riley.[6] The firm became legal advisor to the Supply System, just as it provided legal services for most of the state's public utility districts. Somewhat provincial, steeped in rural and small-town politics and determined champions of public power, those photographed typify the Supply System's roots. They were legitimate heirs to the democratic spirit of the movement for public power in the Northwest.

However, at the same time, the formation of the Washington Public Power Supply System symbolized a different kind of politics. Although Public Utility Districts had been founded by popular vote and their commissioners were elected officials, they had the legal right to borrow money to finance projects without voter approval. WPPSS was one step further removed from popular control. Delegates from member utilities served on

[6] The picture appears in Billington, *People, Politics & Public Power*, 135. Public power in the Northwest was largely a male preserve. When, for instance, the Northwest Public Power Association held its conferences, members were entreated to "Bring the ladies!" A "Ladies' Committee" would prepare a program of tours and social events for those accompanying their husbands. See, e.g., *Pacific Northwest Public Power Bulletin*, 13 (September, 1969): 1.

its Board of Directors and the Board's Executive Committee, but WPPSS itself lacked direct accountability to citizen-ratepayers.

The Supply System not only had the right to borrow money for capital projects, this power – and the tax-free income that its bonds would provide bondholders – was its raison d'etre. States themselves and local governments have legal and constitutional limitations on their borrowing activities – how much debt they can incur and for what purposes. As Annmarie Hauck Walsh points out, the restraints stimulated the use of special authorities for public projects: "Were it not for these encumbrances, public enterprise in states and municipalities would be more frequently undertaken within regular government structures (as it is within the federal government) than in independent authorities."[7] Authorities like the Supply System characteristically issued revenue bonds, payable by the income of the projects they were undertaking, rather than general obligation bonds, repaid by tax proceeds. Revenue bonds usually escaped limits on general obligation municipal borrowing.[8]

In the post–World War II period, state agencies that lacked the direct political and fiscal controls on borrowing by local governments proliferated. By 1961 one author noted that fewer new public authorities were being created. "Authorities may be running out of new things to do," he remarked.[9] Yet his comment preceded a new wave of public authorities created in the 1960s and 1970s. The authorities expanded well beyond traditional public works, raising money to build college dormitories, factories, sports arenas and hospitals, to name a few.[10]

The case of WPPSS indicates how the democratic impulses of the public power movement paradoxically created an organization lacking the accountability that might have prevented it from pursuing nuclear chimeras. In 1968, justifying the selection of WPPSS as the builder of the region's first public power nuclear plant, the Supply System's Managing Director contended that lack of voter control was an organizational asset. "Ability to issue bonds by action of the controlling body and the consent

---

[7] Annmarie Hauck Walsh, *The Public's Business: The Politics and Practices of Government Corporations, A Twentieth Century Fund Study* (paperback ed.; Cambridge and London: The MIT Press, 1980), 23.

[8] For an incisive treatment of the political economy of authority borrowing, see Alberta M. Sbragia, *Debt Wish: Entrepreneurial Cities, U.S. Federalism, and Economic Development* (Pittsburgh: University of Pittsburgh Press, 1996), especially chapters 6 and 7.

[9] Robert Gerwig, "Public Authorities in the United States," *Law and Contemporary Problems* 26, 4 (Autumn 1961): 614.

[10] See Sbragia, *Debt Wish*, passim.

of the parties to the arrangement without referring to a vote of any electorate or other third party veto" was a reason to designate WPPSS for the
task.[11]

Contemplating the paradoxes of democratic governance was no doubt
far from the minds of those who convened at the Washington Public Utility Districts Association office in Seattle on February 20, 1957, for the
organizational meeting of the Supply System. Representatives from sixteen districts were present, along with about a dozen others, mostly those
long associated with the Northwest's public power movement. The gathering needed more space and soon reconvened at a nearby hotel. Along
with operating rules, the group adopted a statement of purposes: "To
construct and operate a system of works, plants, and facilities for the
generating and transmission of electricity described in the order creating this...agency...." More concretely, the Board's newly elected chair,
Grover Greimes, listed agenda items for the first meeting of the Board's
Executive Committee. The last two items were "Major steam plant" and
"Atomic energy generation."[12] The Supply System was new and small,
but it was thinking big.

When the Executive Committee met two weeks later, a rather abstract
discussion of the future of thermal generation transpired, but Jack Cluck
"brought up the subject that private companies are buying acreage near
the Hanford plant. He advised that it be checked to find if the companies
planned to use heat from the Hanford atomic plants. He suggested writing Congressman Jackson for information."[13] The public-private utility
rivalry was never far from the thoughts of Cluck, an ardent liberal committed to the public power movement.

For the short run, getting the organization established and studying
hydroelectric power matters dominated the work of the new Joint Operating Agency. Ken Billington recommended in May that the System hire a
"full time staff man," and in August a committee report favored J. Frank

[11] Owen W. Hurd to Supply System Board of Directors, "BPA Power Purchase Plan Implementing Entity," April 26, 1968. Document in possession of attorney Martha L. Walters,
lent to author.

[12] *Minutes of the Board of Directors and Executive Committee WPPSS 1957–63*, organizational meeting, February 20, 1957, pp. 2, 10–11. (Looseleaf volume at headquarters of
the Supply System, Richland, Washington. References to Board of Directors' minutes will
hereafter be cited as "Minutes, WPPSS Board of Directors" and its Executive Committee's
as "Minutes, WPPSS Executive Committee.")

[13] Minutes, WPPSS Executive Committee, March 8, 1957, pp. 3–4. In fact, Jackson had
moved from the House to the Senate in 1953.

Ward, former head of the Washington State Power Commission, as Managing Director with Owen Hurd, manager of the Benton County PUD, as a second choice. Ward's experience with the unpopular Power Commission had alienated some public utility leaders. Hurd, an experienced engineer himself, was elected by secret ballot. Soon, Hurd was at work, with a salary set at $16,000 a year.[14] In the Northwest's public power circles, Owen Hurd was highly respected, even revered. "Owen Hurd is somewhat a saint in my book," noted the Supply System's long-time General Counsel, Richard Quigley.[15] But Hurd also exemplified the provincial roots of the System.

One effect of Hurd's selection was that the Supply System, at his behest, located its headquarters in Kennewick. This small city (population 14,244 in 1960) adjoined Richland at the edge of the Hanford Nuclear Reservation. Kennewick was the home of Hurd's Benton County PUD, so the choice was a natural one for him. The Board and its Executive Committee continued to hold most of their meetings in Seattle. Yet the headquarters' move to the Hanford vicinity made a nuclear future more likely. The Supply System operated on a daily basis in a "nuclear culture."[16] The annual celebration of Atomic Frontier Days was a variation on the pioneering themes in western community fetes. The football team at Richland's Columbia High School, nicknamed the Bombers, wore mushroom cloud emblems on their uniforms and helmets. The consensus on nuclear development in the Tri-Cities region sheltered WPPSS from the brunt of much of the protest that developed against nuclear energy in later years.

In Ken Billington's pungent phrase, the Supply System had to "get pregnant."[17] By 1957, it was clear that the Federal government was not about to sell off BPA, so WPPSS needed to construct a facility in order to fulfill

[14] Minutes, WPPSS Board of Directors, May 17, 1957, p. 6; August 16, 1957, p. 7; Minutes, WPPSS Executive Committee, September 20, 1957, p. 4. Hurd's salary placed him among the best-paid public utility managers in the region. See, e.g., "Thirteenth Annual Survey of Wages and Working Conditions, September 1959," *Pacific Northwest Public Power Bulletin*, 13 (September 1959): 8.

[15] Interview with Richard Quigley, Kennewick, Washington, October 15, 1992. *Tri-City Herald*, August 18, 1957, p. 1.

[16] The term refers to Paul Loeb's fascinating community study of the Tri-Cities (Richland, Kennewick and Pasco), *Nuclear Culture: Living and Working in the World's Largest Atomic Complex* (paperback ed.; Philadelphia: New Society Publishers, 1986). See also John M. Findlay's intriguing article, "Atomic Frontier Days: Richland, Washington, and the Modern American West," *Journal of the West*, 34, 3 (July 1995): 32–41.

[17] Seligman, 24.

its mission. It spent most of its first year searching for a suitable one. Without a project, the Supply System would be vulnerable to the attacks of the private utilities and conservative politicians. In March 1958, an opportunity appeared. Lewis County PUD had investigated a hydroelectric dam on Packwood Lake in the central Cascades but decided it was too much for the small district to take on by itself. It proposed that WPPSS build the dam. Three weeks later, the Supply System's Board authorized filing a preliminary permit application with the Federal Power Commission and commencing engineering studies. By September 1958, the FPC permit had been issued. The engineering reports were favorable. Power from Packwood could be used to cut peak load demands that WPPSS members put on the Bonneville system.[18]

Packwood Lake Dam made sense to the Supply System. However, it only whetted the appetite of the Supply System and its members for more construction. The project's capacity, about 27.5 megawatts electric (commonly abbreviated MWe or MW), was puny in comparison with the large dams in the Columbia River system and the anticipated size of new nuclear projects. Operating at full capacity, it could only serve a town about the size of Richland (1960 population 23,548).[19] Packwood also pointed up a fact of life for utility planners. Even this modest project took a long time to complete. Launched in 1958, the dam's powerhouse did not go into service until 1964. Yet Packwood Lake's completion remained a point of pride for the Supply System.[20]

More dam building was one possibility. The Ben Franklin site on the mid-Columbia was one of the few places on the river where major hydropower development was possible. More than one utility had cast an eye on the location, which was on the Hanford Reservation. Late in 1958, WPPSS applied to the Atomic Energy Commission for an access permit and security clearance for planning work at Ben Franklin. But the Supply System had more than hydropower on its mind. It also stated that the access would benefit its interest in "the commercial production of electric power from atomic energy."[21] The AEC replied with objections to the possible impact of a Ben Franklin dam on water supplies for its military plutonium production at Hanford. However, its counter-proposal

[18] Minutes, WPPSS Executive Committee, March 21, 1958, p. 2; WPPSS Board of Directors, April 11, 1958, pp. 4–7; WPPSS Board of Directors, May 15, 1959, p. 2.

[19] Population figures in http://www.ofm.wa.gov/pop/decseries/pop189000rev.xls, accessed June 14, 2004.

[20] Billington, *People, Politics & Public Power*, 418.

[21] Minutes, WPPSS Board of Directors, November 21, 1958, p. 3.

for two smaller dams was, according to Hurd, "highly questionable" for WPPSS.[22] In fact, the site remains undeveloped today.

The stymied Ben Franklin project left the Supply System with an unfilled thirst for more activity. Five public utility districts, including Snohomish County, Washington State's largest, decided to drop out of WPPSS in 1958 and early 1959, and the agency's future seemed dim. Packwood Lake had amounted to only a tentative baptism in the power development business. WPPSS sought full immersion. It fought to build a high dam on the Middle Snake River, first against and then in uneasy alliance with the private utilities, from 1960 to 1975, when Congress ultimately protected against dam construction on the Middle Snake under the National Wild and Scenic Rivers Act (see chapter 1). For its efforts over a decade and a half, the Washington Public Power Supply System had amassed substantial legal bills, years of frustration, and no power. Particularly galling to WPPSS was the fact that the Bonneville Power Administration had not supported its proposals; BPA apparently remained unwilling to back a public power project over a private one and unconvinced that a dam was needed at all.[23]

As the dam controversies became more tangled, the nuclear option looked better and better. In the early 1960s, the quickest, most convenient and most politically acceptable route to the Supply System's nuclear rendezvous ran through the Hanford Reservation. Yet even this path, as alluring as it was to the public power community, was not smooth and straight.

The Hanford Generating Project

As far back as 1956, Senator Henry M. Jackson had introduced proposals for a large dual-purpose reactor at Hanford. On the military side, the reactor would irradiate fuel rods to produce plutonium for use in the nation's nuclear weapons programs. Older plutonium reactors at Hanford were nearing the end of their careers. A new production reactor would ensure adequate plutonium for decades of future bomb manufacture. At the same time, Jackson hoped, the reactor's heat could be harnessed for a steam generator to produce electricity for the region. The politics of the dual-purpose reactor fit Jackson's goals neatly. His militaristic

[22] Minutes, WPPSS Executive Committee, January 31, 1958, pp. 2–3.
[23] John Stewart Miller, "Superorganization as an Interorganizational Strategy," PhD diss., University of Oregon, 1977, p. 167.

anti-communism internationally was balanced by his support for activist government at home, especially when it might benefit his state and region. A dual-purpose reactor at Hanford would reverse the direction of the Republicans' partnership doctrine and give the region a new federal generating source.

Despite Scoop Jackson's clout, the dual-purpose reactor faced political obstacles. Coal interests in the East objected to federal support of their competitor, nuclear energy. Congressional Republicans wanted to privatize civilian nuclear power development. Private utilities looked warily on further federal involvement. Moreover, the technology was untested. The other plutonium-producing reactors at Hanford, dating from 1943, had never been used to generate electricity, and a dual-purpose system required extensive design work. Senator Jackson's proposal for the Atomic Energy Commission to build the project died in the House of Representatives in 1956. Two years later, following a classified report to the Joint Committee on Atomic Energy that advocated more plutonium manufacture, President Eisenhower reluctantly signed an Atomic Energy Commission authorization bill that funded a new production reactor at Hanford. Construction began in 1959, and the design included features that would allow later addition of power-generating facilities.[24]

During Kennedy's presidency, the political balance shifted toward support of dual-purpose development. Interior Secretary Stewart L. Udall backed broad-scale power planning for "larger interrelated systems and goals."[25] Five different reports had upheld the technical and economic feasibility of using the new N-Reactor (Hanford reactors were designated by letter) to generate electricity. However, although the Senate allocated funds for conversion to dual-purpose operation in 1961, Republicans, along with a bloc of Democrats primarily from coal-mining states, killed the proposal in the House of Representatives. While some Northwest Republicans had backed the plan, private utilities in the region remained hostile to a federal generating project.[26]

Following the defeat, regional interests began to look for another way to undertake power generation at the N-Reactor. A seemingly ideal instrument for an acceptable proposal was close at hand. "Why don't we build

---

[24] The most comprehensive source on dual-purpose reactor politics in the fifties is Bonnie Baack Pendergrass, "Public Power, Politics, and Technology in the Eisenhower and Kennedy Years: The Hanford Dual-Purpose Reactor Controversy, 1956–1962," PhD diss., University of Washington, 1974.

[25] Cited in Pendergrass, "Public Power, Politics, and Technology...," p. 91.

[26] Haston, "From Conflict Politics," .271.

it ourselves?" said Washington Conservation Department Director Earl Coe to Ken Billington of the Washington PUD Association. "We have just the organization with which to do it," replied Billington.[27] The Washington Public Power Supply System offered many advantages. It could raise capital for the project at favorable interest rates through its authority to issue tax-exempt revenue bonds. The Supply System promised greater financial and organizational strength than a single utility standing alone. WPPSS was fortuitously headquartered in Kennewick, in the Tri-Cities near Hanford. Perhaps most important, WPPSS was eager. Looking for greater visibility and a larger organizational role in Northwest power politics, the Supply System's leaders wanted the job.[28]

The decision to designate WPPSS to finance, construct, and operate the generating facilities at Hanford was one of two key choices that turned the project's 1962 congressional defeat into victory a year later. The second was Henry Jackson's sudden proposal to offer half of the N-Reactor's power output to the region's private utilities. This gesture, which Jackson later claimed "came right off the top of my head," converted the companies from opposition to still-wary acceptance of the project. Public power diehards found this a "bitter pill," but swallowed it anyway.[29]

Jackson's proposal gave the Atomic Energy Commission authority to sell the steam produced in the plutonium manufacturing process to the Supply System. WPPSS would finance, build, and operate a generating plant to convert the steam into electricity. Although dual-purpose reactors had been built in the Soviet Union, the Hanford project was unique and untested in the West. Scientists and engineers were even at first uncertain of the capacity of the generating facility. It was rated at 800 megawatts, which would make it by far the largest source of nuclear power in the United States when completed and operating.

To integrate the electricity into the regional network, the N-Reactor plan proposed a complex scheme. Utilities in the region would commit to shares of the plant's generating capability, with half available for the private companies and half for public utilities. WPPSS, however, would supply the power to the Bonneville Power Administration, which would in exchange deliver an equivalent dollar amount of power to the participating utilities. Since the Generating Project's thermal power would

---

[27] Billington, *People, Politics & Public Power*, 183.
[28] Pendergrass, "Public Power, Politics, and Technology...," 129.
[29] Pendergrass, "Public Power, Politics, and Technology," 136; Billington, *People, Politics & Public Power*, 184. See also Haston, "From Conflict Politics...," 270–274.

cost more per kilowatt than electricity from the BPA's hydropower system, the exchange benefitted Hanford project participants. For example, if BPA received energy costing six mills (0.6 cents) per kilowatt-hour from the N-Reactor and exchanged it for its two mills per kilowatt-hour electricity, the participants would receive three times as many kilowatt-hours as the plant would generate. The Hanford Generating Project exchange arrangement foreshadowed marketing methods for the first three of the Supply System's own nuclear reactor projects. It established the principle that utilities and their ratepayers would pay for new energy sources not at their marginal costs but at the average cost to the regional system. Although in the early 1960s enthusiasts expected nuclear power costs to plunge, the gap between nuclear and hydropower costs meant that the marketing arrangement subsidized nuclear energy.[30]

At the beginning, however, the Hanford Generating Project looked like a great bargain to Supply System leaders. In a 1963 speech, Owen Hurd predicted that dual-purpose operation would produce electricity for a mere 2.5 mills per kilowatt-hour. Without it, Bonneville faced shortages for its preference customers by 1965; the project would, he maintained, provide enough energy to meet more than two years of anticipated load growth for Bonneville's preference customers. Moreover, the exchange agreement offered a model of marketing electricity that could apply to other new projects.[31]

Congress approved Senator Jackson's revised proposal and President Kennedy signed it into law on September 26, 1962. The plutonium production facility was already nearing completion, and conversion to dual-purpose operation began exactly one year later. The President broke ground for the Hanford Generating Project before a crowd of 37,000. For the first time, authorities allowed the public to enter the Hanford Reservation for the ceremony.[32]

The N-Reactor project was emblematic of the symbiosis of the Cold War with the mid-century American boom economy. At the groundbreaking,

[30] Lewis Strauss speech of September 16, 1954 quoted in Clarfield and Wiecek, *Nuclear America*, 277.
[31] Owen W. Hurd, "Regional Benefits of Hanford New Production Reactor Power Facilities," speech to First Electro-Nuclear Conference, Institute of Electrical and Electronic Engineers," Richland, Washington, April 30, 1963.
[32] Pendergrass, "Public Power, Politics, and Technology...," 175. Poignantly, the Supply System's minutes of a meeting on the day of President Kennedy's assassination report that "35 requests have already been received for showings of the new color-sound movie" of his groundbreaking visit. See Minutes, WPPSS Board of Directors, November 22, 1963, p. 2.

Kennedy observed, "It was appropriate to come here where so much has been done to build the military strength of the United States and to find a chance to strike a blow for peace and for a better life for our fellow citizens."[33] President Eisenhower had considered the new plutonium reactor unnecessary in part because his administration's strategy of massive retaliation depended on a relatively small arsenal of hydrogen bombs. The Kennedy administration shifted toward more "flexible" military strategies and inclined toward a "counterforce" doctrine. These plans to wipe out Soviet weaponry rather than threaten its population centers entailed a buildup of nuclear warheads and delivery systems, and hence continued plutonium production.[34] The Hanford Generating Project itself fed the energy appetites of Northwestern homes, farms, and stores but also the military operations on the Hanford Reservation itself, the aluminum companies which had located in the Northwest, and Boeing (the nation's second-largest military contractor in fiscal year 1963).[35]

For the Washington Public Power Supply System, the Hanford Project brought valuable positive publicity. "Today," wrote Owen Hurd on the eve of its opening, "the eyes of the Northwest, and the entire Nation are on Hanford and the world's largest nuclear steam plant. All connected with public power can be rightfully proud...."[36] Hanford gave the System its first big, visible task. It provided experience in large-scale project management. It made WPPSS a focus of attention for those considering how to launch more major generating facilities. In its ties to the military, its links with the BPA, and its arrangements to market power to public and private utilities alike, WPPSS situated itself as a player in the Northwest's version of the growth-oriented mixed economy of Cold War America.[37]

---

[33] Quoted in Tollefson, *BPA and the Struggle*, 333.

[34] Clarfield and Wiecek, *Nuclear America*, chapter 9 provides a good summary of Kennedy-era nuclear strategy.

[35] Boeing had 5.2 percent of Defense Department prime contracts for the year. See "Department of Defense Lists Top 100 Prime Contractors for Fiscal 1963," *Aviation Week and Space Technology*, December 30, 1963, 60.

[36] Owen W. Hurd, "World's Largest Nuclear Plant Nears Operation," *Public Power*, 24, 4 (April 1966): 16–18, quoted at p. 18.

[37] Some of the treatments of the broader political economy which have influenced my thinking are: Louis Galambos and Joseph Pratt, *The Rise of the Corporate Commonwealth: U.S. Business and Public Policy in the Twentieth Century* (New York: Basic Books, 1988); Herbert Stein, *The Fiscal Revolution in America* (Chicago and London: University of Chicago Press, 1969); Stephen A. Marglin and Juliet B. Schor, eds., *The Golden Age of Capitalism: Reinterpreting the Postwar Experience* (Oxford and New York: Clarendon Press Oxford University Press, 1990); Samuel Bowles, David M. Gordon, and Thomas E. Weisskopf, *Beyond the Waste Land* (paperback ed.; Garden City, NY: Doubleday Anchor

On the other hand, the N-Reactor project bestowed problems as well as opportunities on the Supply System. In the fall of 1962, when Senator Jackson's compromise was enacted, the System's Executive Committee wanted "to move ahead rapidly" to meet a 1965 target for power production.[38] But grafting a turbine-generator onto a plutonium reactor proved a tricky job. Initial cost estimates of $150 million had grown to about $190 million by groundbreaking. Commercial operation did not begin until April 1966.[39]

The technology of the dual-purpose reactor proved flawed. Breakdowns plagued the Hanford Generating Project and burdened the BPA network with its unreliability. From the standpoint of the public utilities, Bonneville's hydro preference customers, the N-Reactor created special difficulties. Participants, public and private, exchanged whatever power the facility produced for an equivalent dollar amount of BPA power. The investor-owned utilities (IOUs) had nicked a small hole in the preference principle, for they shared equally with the public customers in obtaining this low-cost supply. When the N-Reactor project was not operating, BPA was still mandated to deliver the equivalent of its capability to participating private utilities on the same terms as the publics. To Ken Billington, the Hanford Project arrangement was a bad deal for preference customers. In 1974, he described the Hanford arrangement as a horse and rabbit stew: private utilities "got the horse, while we got the rabbit."[40]

The Hydro-Thermal Power Program

In the mid-1960s, the travails of the N-Reactor generating project were in the future. Nationally, nuclear electrical generation was beginning its take-off. The Pacific Northwest's nuclear bandwagon also began to roll. Under the aegis of the Bonneville Power Administration, public and private utilities commenced planning for new power supplies to meet expectations of regional growth. As we have seen, proposals to supplement hydropower with thermal generation had been floated since the early 1950s, but now ideas began to crystallize as plans. Charles Luce, who

Books, 1984); Michael J. Piore and Charles F. Sabel, *The Second Industrial Divide: Possibilities for Prosperity* (New York: Basic Books, 1984).

[38] Minutes, WPPSS Executive Committee, September 21, 1962, p. 3.

[39] Pendergrass, "Public Power, Politics, and Technology...," 176.

[40] Pendergrass, "Public Power, Politics, and Technology...," 182–183; Billington, *People, Politics & Public Power*, 184–185, 316–317.

had espoused nuclear development throughout his tenure as Bonneville Administrator, left in August 1966, but his successor, David S. Black, adopted Luce's goal. On October 13, 1966, he convened a meeting of regional utility officials and appointed a task force to plan for thermal power generation. A month later, the group was expanded and formalized as the Joint Power Planning Council (JPPC). Representatives of 104 public and four private utilities, along with BPA staff and delegates from other groups with interests in energy planning, comprised the council; Bonneville's Administrator chaired it.

Although cooperation between public and private utilities was a watchword of the JPPC, the public agencies viewed the inauguration of formal planning as a signal to consolidate their own forces. It was not that they lacked enthusiasm for nuclear energy. The Northwest Public Power Association, hearing reports of the impending regional planning effort, agreed in October 1966 to make its next convention theme "Entering the Nuclear Age." Nor were they opposed to comprehensive planning. Alex Radin, longtime General Manager of the American Public Power Association, had advised the region to form a "regional agency," to "build big" and to "build together."[41] Thus, public power shared the regional faith in thermal energy and the region's commitment to economic growth.

However, public utilities had been notoriously divided themselves. The larger publics with their own generating facilities had sometimes cast their lot with the investor-owned firms. Smaller publics, dependent on BPA for all of their electricity, felt the need for a group to express public power interests in the planning process. They spurred formation of the Public Power Council (PPC), headquartered in Vancouver, Washington, and comprised of all publicly owned utilities within the BPA service area.[42] According to Alan Jones of McMinnville, Oregon, the first chairman of the PPC, "Without a Public Power Council, investor-owned utilities and the larger customer-owned utilities would dominate power planning in the region."[43]

At times the Joint Power Planning Council was contentious, but there was little dispute about the fundamentals of its plan: The well-being of the Northwest's economy depended on abundant, low-cost electricity. Increasing power demand would necessitate aggressive expansion of

---

[41] Cited in Miller, "Superorganization," 130, 124.
[42] Billington, *People, Politics & Public Power*, 237–239, describes the delicate negotiations over creation of the PPC.
[43] Miller, "Superorganization," 135.

regional supply. Without major new dam sites, thermal plants would be the bulwarks of the new plan. Although some coal-fired generators could serve the Northwest, the cleanest and cheapest future power source would be nuclear. Nuclear plants, in turn, would have to be large in order to capture economies of scale. To serve the spectrum of regional needs, these plants should operate on a coordinated basis with each other and with the hydropower system. The cost of thermal power should be blended with inexpensive hydropower in Bonneville's rates. Plants should be geographically dispersed to supply different parts of the Northwest.

Finally, a mixture of public and private ownership arrangements would be necessary. As an era of shortage seemed to draw near, all utilities wanted to control their own generating resources. Gus Norwood, former head of the Northwest Public Power Association, had spelled it out in a speech to public power groups in 1965: "Putting it bluntly, the lesson...is to get your own power supply or a very friendly power supply or in the alternative, fast or slow, your electric system will die."[44] No utility or group of utilities wanted to be caught depending entirely on someone else's generating plants.

Demand forecasts reinforced the assumptions of Northwest energy planners. Since the early 1950s, the Pacific Northwest Utilities Conference Committee (PNUCC), a group representing the region's utilities, had issued an annual twenty-year load and resource forecast covering the so-called West Group area – roughly Washington, Oregon, northern Idaho, and Montana west of the Continental Divide. Yet the PNUCC itself did not prepare the predictions it published. It simply added up the forecasts it received from more than one hundred utilities serving the West Group territory. In turn, smaller public utilities relied on BPA staff to prepare their forecasts. Bonneville also calculated the loads of the Direct Service Industries, which received their energy from BPA under long-term contracts. In sum, Bonneville estimates accounted directly for about 40 percent of the demand projections in the PNUCC's report.[45]

---

[44] Gus Norwood Papers, University of Washington Manuscripts and Archives Division, Box 2, Folder "Norwood-Speech-May 6, 1965, APPA Panel, Los Angeles, 'Power Supply – Live or Die'."

[45] U. S. General Accounting Office, *Bonneville Power Administration and Rural Electrification Administration Actions and Activities Affecting Utility Participation in Washington Public Power Supply System Plants 4 and 5* (Washington, DC: 1982), 4. The BPA-PNUCC connection is described and criticized in Ernst & Ernst, *Review of Energy Forecasting Methodologies and Assumptions*, prepared for Bonneville Power Administration, U.S. Department of Interior, June 1976, section III.

Given even a modest degree of hindsight, it is not hard to find flaws in PNUCC-BPA methodologies. The sum-of-utilities forecast simply accumulated the biases of the individual utility predictions. Thus, a local utility hoping to attract an employer, developer, or commercial customer to its service area would understandably project a need for the energy to serve the potential load. So might a utility in a neighboring community, competing for the same customer. The PNUCC forecast had no way to eliminate the tendency toward counting – and sometimes double-counting – the aspirations instead of the realistic expectations of local utilities. The four large Northwest private utilities, as well as most of the major publics, did conduct their own forecasts. As late as 1976, these ranged in sophistication from fairly elaborate econometric models to Pacific Power & Light's method of assuming a constant annual growth rate after correcting for weather conditions.

The technical deficiencies of Northwest energy forecasts reflected an underlying cultural reality. Simply put, electricity was an unalloyed good – the more of it, the better. The *Northwest Ruralite*, published for members of the region's rural electrical cooperatives, commented in 1962: "electric power is like love – nobody ever gets quite enough."[46] Electricity had qualities beyond the utilitarian. It added splendor as well as ease to people's lives. Electricity was both mundanely practical and magically transformative.[47] Northwesterners, recent beneficiaries of widespread electrification, took it for granted that they would want all the cheap electricity they could get. Nevertheless, in the 1950s and 1960s, and indeed through about 1973, the Bonneville-PNUCC forecasts had one virtue that outweighed all deficiencies. They accurately predicted the rapid growth of electrical energy usage in the Pacific Northwest. Successful forecasting justified planning for rapid growth. Rapid growth in turn gave credence to forecasts that it would continue indefinitely. When reality began to diverge from earlier forecasts, energy planners' adjustments were slow and grudging, as Table 2.1 indicates.

The Washington Public Power Supply System, having undertaken the Hanford Generating Project, wanted to lead public power's nuclear development. While building the Generating Project, it was simultaneously

---

[46] Quoted in Pendergrass, "Public Power, Politics, and Technology," 156–157.

[47] For discussions of the multiple meanings of electricity, see Carolyn Marvin, *When Old Technologies Were New: Thinking about Electric Communication in the Late Nineteenth Century* (New York: Oxford University Press, 1988); David E. Nye, *Electrifying America: Social Meanings of a New Technology* (Cambridge: MIT Press, 1990).

TABLE 2.1. *Demand growth rates: forecast and actual, 1969–1980*

Forecast date	1969/70–1973/74	1973/74–1976–77	1976/77–1979/80	1969/70–1979/80	1973/74–1979/80
1969	6.06%	5.38%	5.56%	5.70%	
1973		5.26%	5.24%		5.25%
1976			5.88%		
Actual Load Growth	3.75%	4.35%	3.78%	3.94%	4.07%

	1973/74	1976/77	1979/80
1969 Forecast Overestimate	4.8%	12.0%	17.9%

Figures are compound annual growth rates for intervals indicated. In last row, figures represent (predicted – actual load)/actual load.

Source: Adapted from Office of Applied Energy Studies, Washington Energy Research Center, Washington State University/University of Washington, *Independent Review of Washington Public Power Supply System Nuclear Plants 4 and 5: Final Report to the Washington State Legislature* (n.p.: Washington Energy Research Center, 1982), p. 77, Table 3.1.

investigating light water reactors, already becoming the nation's standard type, and an alternative model known as a Heavy Water Moderated Organic Cooled Reactor.[48] By the end of 1965, the Supply System decided to write to the Atomic Energy Commission to stake a claim for possible involvement in the heavy water project. The Board also agreed to devote at least one employee to the project. Soon, Supply System staff began meeting with the companies and Bonneville representatives about the program.[49]

WPPSS continued to plan on a variety of fronts. It brought in Kenneth A. Roe of Burns & Roe, a leading engineering consultant, to discuss site selection. Roe and the Supply System's Directors agreed they needed to choose sites for nuclear plants soon. Intimating conflicts that emerged later, Roe "stressed... the importance of public acceptance... and also pointed out that the Hanford area is one of the best in the country due to numerous reasons: public acceptance, location away from population areas, no earthquake danger, plenty of cooling water, etc."[50]

[48] Cowan, "Technological Lock-in," provides analysis of the industry's commitment to light water reactors. See also Morone and Woodhouse, *The Demise of Nuclear Energy?*, chapter 2.
[49] Minutes, WPPSS Board of Directors, December 3, 1965, p. 5; Minutes, WPPSS Executive Committee, May 20, 1966, p. 2.
[50] Minutes, WPPSS Board of Directors, June 24, 1966, p. 5.

Thus, even before the Hydro-Thermal Power Program, WPPSS was eager to act. In September 1966, Owen Hurd sent Bonneville a preliminary plan for a 1,000 MWe light water nuclear plant to be built at the Hanford Reservation, proposing at the same time that a demonstration reactor with a heavy water design also be considered.[51] Then, at a January 1967 meeting in Portland, Hurd outlined a proposal for a regional plan. The document he presented carefully noted that the Supply System's Board of Directors had endorsed the presentation, that the Public Power Council's Executive Council had "encouraged" it, and that it came as a response to a request from Bonneville Administrator Black. Since the proposal anticipated many of the policies later adopted for the first three WPPSS nuclear plants, it merits some detailed examination.

Hurd began with a surprising admission. The date when BPA would no longer be able to supply enough firm energy to its preference customers had "always been a receding target." The current estimate of 1980 might itself be too early. Yet he quickly set this aside and called for prompt action to meet four objectives, spelled out in capital letters:

1. PROVIDE AMPLE POWER AT LOWEST COST TO MAINTAIN THE REGION'S COMPETITIVE POSITION IN SERVING LARGE INDUSTRIAL LOADS
2. PROVIDE POWER TO MEET THE LOAD GROWTH OF ALL UTILITIES
3. POSTPONE BPA FIRM POWER INSUFFICIENCY
4. EQUITABLE SHARING OF BENEFITS AND RISKS

Bonneville could not solve these problems alone. Without new authority from Congress, the BPA could neither build new power plants itself nor directly purchase electricity from non-federal plants. The Washington Public Power Supply System was the right vehicle "for securing maximum benefits resulting from joint action in the field of power generation and transmission."[52]

---

[51] Owen W. Hurd to H. R. Richmond, September 14, 1966. Document in possession of attorney Martha L. Walters, lent to author.

[52] "A Proposed Plan for Northwest Thermal Plants to Meet Regional Action Needs and Objectives by BPA and Utility Cooperative Action," presented by Owen W. Hurd, Managing Director, Washington Public Power Supply System, at BPA Thermal Task Force Meeting, Portland, January 23, 1967. Plaintiff's Exhibit 177, DeFazio v. WPPSS, Lane County Circuit Court, 1974. (These exhibit documents lent to author by Martha L. Walters, an attorney in the case.)

Hurd went on to propose that WPPSS build a 1,000 megawatt light water reactor nuclear plant on the Hanford Reservation. The Supply System, by quickly securing a reactor and turbine generators, could have the plant ready for commercial operation by July 1972. Cost calculations favored this plan. A nuclear system would cost about $123 million to build, slightly more than a coal plant, but fuel expenses would be much lower. The total cost would depend on the plant factor (the percentage of time that it would operate), but if, as Hurd expected, the nuclear plant could operate more than 70 percent of the time, it would produce electricity more cheaply than coal. Again depending on plant factor, costs would range from 2.8 to 4 mills per kilowatt-hour, only slightly more expensive than federal hydropower.[53]

The WPPSS proposal became public power's entree into planning for nuclear power in 1967–68, even though the Supply System itself drew some fire from other public utility interests. Billington fretted in October 1966, "WPPSS now appears to have its mind made up on where, when and how the next plant is to be built. This is not immediately acceptable to some of our generating utilities...."[54] However, not ready to put aside their suspicion of private power, managers of Washington Public Utility Districts voiced "strong sentiment for using WPPSS as public power's power supply arm.... [I]n the face of refusal of the private companies to work with WPPSS perhaps the Public Power Council should say it has to be 'WPPSS or else.'"[55] Meanwhile, the Supply System pushed ahead and ordered a turbine-generator from Westinghouse and a nuclear steam supply system (a reactor vessel and related equipment) from General Electric. Expected commercial operation slipped back to "sometime in 1973."[56]

Throughout the deliberations of the Joint Power Planning Council, public and private utilities jockeyed for position. Two investor-owned utilities planned large coal-burning plants outside Centralia, Washington, near the largest coal fields in the region. But this coal was of fairly low quality and hard to mine.[57] At the same time, Portland General Electric announced

---

[53] Ibid.

[54] Billington, *People, Politics & Public Power*, 235.

[55] Minutes, Managers' Section, Washington Public Utility Districts Association, March 23–24, 1967, in Box 12a, Washington Public Utility Districts Association papers, Special Collections, University of Washington Library, Seattle.

[56] Minutes, Washington Public Utility Districts Association, May 19, 1967, Box 12a, Special Collections, University of Washington.

[57] Endel J. Kolde, *From Mine to Market – A Study of Production, Marketing and Consumption of Coal in the Pacific Northwest*, Occasional Paper Number 3 (Seattle: University

it would build a nuclear facility, named Trojan, along the Columbia at Rainier, Oregon. The Eugene Water and Electric Board (EWEB) agreed to take a 30 percent ownership share of the Trojan plant. Washington public power interests saw this as an example of EWEB's tendency to ally with private utilities. In fact, by March 1968, Ken Billington reported that "many public power people" felt "frustrated [that] the private companies are moving ahead with the ball and could control the nuclear power field, if the public agencies do not also move ahead."[58]

Inventing Net Billing

While public and private utilities mixed pieties about cooperation with efforts to outflank one another, Bonneville Power Administration was hard at work on a scheme for financing thermal plants and marketing the electricity they would generate. Bonneville could not own generating facilities nor buy power directly from non-federal plants. Bernard Goldhammer, Bonneville's Power Manager, sought a way to work around these limitations and give BPA a coordinating role in paying for the plants and distributing their output. A decade later, Goldhammer pointed out a paradoxical aspect of the region's energy planning. On the one hand, individual utilities and groups of utilities wanted their own generating facilities rather than dependence on rivals and strangers. Yet, at the same time, they (and especially Bonneville itself) considered centralization and integrated operations as regional necessities.[59] Despite the drive for autonomy that characterized public power in the Pacific Northwest, the seldom-questioned faith in supply-side solutions and economies of scale meant that centralization would likely triumph over localism.

Many considered Bernard Goldhammer a bureaucratic and managerial wizard for his successes in working out the complex arrangements for projects like the Canadian Treaty and the Intertie. His contributions to the Hydro-Thermal Power Program of 1968 were fundamental, especially his scheme for what came to be known as net billing. Under net billing, a utility participating in a thermal power plant would agree to turn over its share of the "capability" of the plant to Bonneville. BPA, which charged its customers monthly for the power it sold them, would in return deduct

of Washington College of Business Administration Bureau of Business Research, May 1956), 7–11, 20–21.

[58] Minutes, Washington Public Utility Districts Association Board, March 15, 1968, p. 3.

[59] Lee, et al., *Electric Power and the Future*, 70.

from its bill to the utility an amount equal to the utility's share of the power it received from the plant. Net billing was a close relative of the exchange agreements the Hanford Generating Project had employed. Under the exchange agreements, Bonneville took the project's output and provided utilities with electricity with the same dollar value. Under net billing, BPA compensated utilities for the electricity from a nuclear plant by a credit on their bill. Bonneville's description some years later is one of the clearest explanations of the complex procedure:

Under this arrangement, preference utilities (public bodies and cooperatives) have built and are building portions or all of certain thermal powerplants to meet their future power requirements. They furnish the output to BPA. BPA in turn bears the preference customers' shares of the costs of those powerplants, acquires the power output, and blends it with Federal hydropower. BPA then sells the blended product to its various customers, including the participating preference utilities. It "pays" those utilities for their shares of the powerplants' costs by reducing their annual bills for power purchases and other services from BPA.[60]

The fancy footwork of net billing had significant implications. First, Bonneville would avoid legal restrictions on its action by receiving the electricity from the thermal plant without purchasing it directly and without owning the facility. Second, net billing would affect the price structure for the region's power supply. Under it, utilities would assign expensive kilowatt-hours from new coal or nuclear plants to the federal agency. BPA then would compute an overall rate that melded this high-priced electricity with cheaper power from the hydro system and would sell the mixture of thermal and hydro at the blended rate to its utility customers. The greater the gap between hydro and thermal power costs, the greater the impact on Bonneville's rates for all its customers. As an example, suppose public utility ABC's share of a net-billed plant's capability equals ten million kilowatt-hours per month, at a cost of 4 cents per kilowatt-hour. At the same time, the utility purchases forty million kilowatt-hours at 2.5 cents per kilowatt-hour from Bonneville. ABC pays the operator of the net-billed plant $400,000, but its share of the energy is delivered to BPA. BPA in turn provides its forty million kilowatt-hours to the utility

---

[60] U.S. Department of Interior, Bonneville Power Administration, *The Role of the Bonneville Power Administration in the Pacific Northwest Power Supply System Including Its Participation in the Hydro-Thermal Power Program*, Final Environmental Impact Statement, December 1980, DOE/EIS-0066, p. I–17.

and deducts $400,000 from its gross bill of $1,000,000 for a net bill of $600,000.[61]

To a utility participating in a new thermal plant, the net billing arrangement offered assurance that its commitment to expanding its power supply would not force it to pay for expensive new thermal power while other utilities in the region were buying low-cost hydro. Economists might note that by pricing new electricity well below its actual marginal cost, Bonneville's net billing arrangements encouraged excessive consumption, but in the late 1960s and early 1970s anticipated efficiency gains from newer, bigger, and cheaper nuclear reactors meant that few expected that the market distortion would be severe. In fact, Northwest planners seldom if ever thought about crucial variables such as the price elasticity of demand for electricity. For example, during a "Dialogue on Power Demands" in 1972, Oregon's State Engineer asked a Bonneville manager, "[W]ill the increasing cost of electricity cut back in the demand for electricity?" The reply: "Well, off-hand I can say we haven't considered that."[62] The price of such inattention would soon prove exorbitant.

There is an ominous hint of another kind of problem in an exchange in early 1969 between Ken Billington and Bonneville Power Administrator H. R. (Russ) Richmond. What would happen, Billington asked, if, in Richmond's words, "an accepted and approved plant fails?" Richmond replied that a failed project would be a regional responsibility for BPA customers: "[W]e would propose that the sunk cost in such an unsuccessful project be net billed commencing immediately."[63] In 1982 and 1983 this eventuality became a reality. WPPSS suspended construction on two net-billed plants; they were terminated in 1994. Bonneville's rates must still generate enough revenue to pay off almost four billion dollars of outstanding bonds for these ventures that never have and never will produce any electricity.[64]

At the time of the Hydro-Thermal Program, however, the danger signals flickered only faintly on the horizon. A more immediate implication of

---

[61] This hypothetical example is similar to one in Lee et al., *Electric Power and the Future*, which has an excellent discussion of net billing, 75–82.

[62] Oregon Nuclear and Thermal Energy Council, *Dialogue on Power Demands* (Salem, OR: The Council, 1972), II–4.

[63] H. R. Richmond to Ken Billington, April 18, 1969, in Box 8, Washington PUD Association Collection. (Cf. also Billington to Richmond, April 14, 1969).

[64] The $4 billion estimate is from Energy Northwest, *2005 Annual Report*, 21–24, at http://www.energy-northwest.com/downloads/FY05.pdf, accessed July 14, 2006.

net billing related to the role of Bonneville's industrial customers. Under the long-standing preference policy, public utilities had first claim on the output of the federal hydropower system. Since before Pearl Harbor, however, BPA had also sold electricity directly to aluminum companies and other electro-process firms. From the days of J. D. Ross, Bonneville had sought these Direct Service Industries (DSIs) to improve its load factor and, more generally, to stimulate regional economic development. Since aluminum plants operated around the clock, they took power generated at times of low residential and commercial demand. In 1965, new industrial sales contracts offered another advantage to the system. In exchange for lower rates, the companies accepted a category of energy called "Modified Firm." Bonneville could interrupt up to one-quarter of the Modified Firm load at any time and for any reason. Thus, that quartile of the DSI companies' normal load became in effect a reserve for the rest of the system when demand was high or facilities had to be shut down for technical reasons.

In 1967–68, Bonneville delivered about 43% of its power to the Direct Service Industries.[65] As long as the preference customers' demand on the hydropower system left BPA with excess capacity, industrial sales were a boon to all. However, the public utilities were beginning to wonder whether they should build high-cost thermal plants and provide the electricity to Bonneville through net billing so that BPA could supply new DSI customer loads. Under the Hydro-Thermal Plan, net-billed plants would not only supply BPA's preference customers but would also provide electricity for the Direct Service Industries' load growth. Public utilities and their customers in effect were to build new power plants so that industrial firms could continue to share the output of the system. Public power was quite aware that its undertakings would also benefit the Direct Service Industries. In 1967, even before the Hydro-Thermal Program incorporated net billing, the Public Power Council recommended that BPA serve new loads of existing DSIs only if they were "unique or experimental."[66] Preference utilities wanted new capacity for themselves, not DSI customers. Public power continued to be wary of Bonneville commitments to serve new industrial loads. One PUD Commissioner, for instance, complained to Bonneville that the agency had ignored the preference

[65] U.S. Department of the Interior, Bonneville Power Administration, *Federal Columbia River Power System 1968 Annual Report*, 28.
[66] Public Power Council Executive Committee policy adopted July 28, 1967, in Box 7, Washington PUD Association Collection.

customers' objections to new industrial sales. The growth requirements of utilities participating in net billing, he insisted, should take precedence over expanding DSI loads.[67]

By October 22, 1968, the Joint Power Planning Council had completed its study and announced its plan. Its report, *A Ten Year Hydro-Thermal Power Program for the Pacific Northwest*, was published in January 1969. The Program was nothing if not ambitious. Although detailed projections concentrated on the 1970s, it set forth what amounted to a twenty-year program. By 1990, they maintained, the Pacific Northwest would need "the equivalent of twenty nuclear plants of 1,000 mw [megawatts] each and one coal-fired plant of 1,400 mw."[68] These would be base load plants, designed to run continuously to meet normal demand. They would be the responsibility of public and private utilities in the region, but Bonneville would acquire the output of the public plants through net billing and could make arrangements with private plants to obtain surplus power from them through short-term exchange agreements. The federal government's direct investment in the program would come in expanding the hydro system. It would complete generating projects already underway or authorized and add another twenty-one projects not yet approved by Congress. Hydro improvements aimed at increasing peaking capacity, the system's ability to handle maximum loads. The federal contribution would also entail a major expansion of the region's transmission grid that Bonneville operated. The switch to thermal base load plants would reverse the previous roles of Northwest electricity sources. By 1990, according to the Hydro-Thermal Program, more than half of the region's firm energy would come from nuclear and coal-fired plants.[69]

The Hydro-Thermal Program would have approximately tripled the electrical energy resources of the Pacific Northwest in a twenty-year span. Average energy use was projected to grow about 5.4 percent per year; peak demand about 5.9 percent annually. The price tag would be substantial, nearly $18 billion between 1969 and 1990. Table 2.2 summarizes the projects, capacity increases, and investment costs foreseen in the plan.

The Hydro-Thermal Power Program's boldness reflected national hopes in the late 1960s for the rapid development of nuclear energy. The White

[67] Public Power Council, Minutes, December 1, 1970. O. G. Hittle, General Manager, Cowlitz County PUD, to H. R. Richmond, January 22, 1970, in Box 8, Washington PUD Association Collection.

[68] Bonneville Power Administration, *Hydro-Thermal Power Program*, 29.

[69] Ibid., 20.

TABLE 2.2. *Hydro-thermal power program – planned projects, capacity, and cost 1969–1990*

Year of Operation	Number of Projects	Capacity (megawatts)	Type
New Thermal Generation			
1971–1980			
Ownership			
Public	0	n.a.	n.a.
Private	5	5,000	Nuclear
Joint	2	1,400	Coal
1981–1990			
Public	5	5,000	Nuclear
Private	10	10,000	Nuclear
Subtotal	22	21,400	Thermal
New Hydro Capacity			
1969–1980		11,671	
1981–1990		10,030	
Subtotal		21,701	Hydro
Total New Capacity		42,101	

Estimated Investment Required	Federal	Non-federal
Thermal Generation		$4,000,000,000
Hydro Capacity	$3,700,000,000	$1,400,000,000
Transmission and Distribution	$2,400,000,000	$6,400,000,000
Total Estimated Investment	**$6,100,000,000**	**$11,800,000,000**

Source: U.S. Department of the Interior, Bonneville Power Administration, A Ten Year Hydro-Thermal Power Program for the Pacific Northwest (January 1969), 30–35.

House Office of Science and Technology in 1968 recommended construction of 250 "mammoth-sized" new power plants by 1990.[70] Some environmentalists, including many members of the Sierra Club, saw properly sited nuclear plants as less destructive than fossil-fuel generation or large dams on scenic rivers.[71] Few people had thought deeply about alternatives to large-scale central-station generation to meet the demand they assumed would grow rapidly and inexorably.

[70] U.S. Office of Science and Technology, Energy Policy Staff, *Considerations Affecting Steam Power Plant Site Selection* (Washington, DC: Executive Office of the President, Office of Science and Technology, 1968), viii.

[71] The national Sierra Club did not call for a moratorium on all new nuclear power until 1974. See Michael P. Cohen, *The History of the Sierra Club 1892–1970* (San Francisco: Sierra Club Books, 1988), 379–80, 385–387 and *passim.*

The Supply System Begins Nuclear Planning

Although it received little public attention, Northwestern utility interests knew the Hydro-Thermal Program marked a turning point in the region's energy policies. "October of 1968 may well have been the most significant month in the history of the Pacific Northwest power industry" since the late 1930s, observed the *Northwest Public Power Bulletin*.[72] New technologies, new managerial structures, and new financial arrangements would henceforth shape the region's efforts.

For the Supply System, the task was to convert planning activities into prompt action. Owen Hurd was certain the region required "large-scale thermal plants...on line at the time needed."[73] Bonneville also wanted rapid progress. Administrator Richmond exhorted the Public Power Council: "At least three of the first seven thermal plants should be publicly owned" and net billed. "Bonneville therefore needs three public power entities to come forth immediately, each agreeing to finance and build a nuclear powerplant on a business-like basis."[74] Without these commitments, the late 1970s would be a time of shortage and crisis. Early in the spring of 1969, WPPSS told the Public Power Council it was ready to act.[75] In the following months, the Supply System and Bonneville concentrated on two tasks, devising the complicated contractual relations among WPPSS, BPA, and the region's preference customers and finding an acceptable reactor location. Both proved to be sticky problems.

Although net billing circumvented the task of securing new statutory authority for Bonneville, the Hydro-Thermal Power Program was nevertheless momentous enough to require consideration in the nation's capital. The prospect of financing expansion with tax-exempt municipal bonds disturbed private power interests. Some politicians in other regions also resented Bonneville's fiscal arrangements that had brought the Pacific Northwest cheap electricity for three decades. It took nearly a year after

---

[72] Henry G. Curtis, "Exec. Secretary's Comments," *Northwest Public Power Bulletin*, 22, 11 (November 1968): 2.

[73] "Northwest Utilities, Bonneville Agree on $15 Billion Joint Hydro-Thermal Power Program," *Northwest Public Power Bulletin*, 22, 11 (November 1968): 3.

[74] "Summary of Report on Implementation of Hydro-Thermal Power Plan," p. 1, Public Power Council Executive Committee meeting, January 22, 1969, Box 8, Washington PUD Association Collection.

[75] Minutes, Public Power Council annual meeting, March 27, 1969, p. 2, Box 8, Washington PUD Association Collection.

the program's unveiling before Richard Nixon's Secretary of the Interior, Walter Hickel, announced that the administration favored the proposal to acquire project capability through net billing and to blend hydro and thermal costs.[76] The legal justification for net billing rested on advisory opinions from the Solicitor in the Interior Department and from counsel in the General Accounting Office, as well as a congressional subcommittee chairman's letter and statements made during the U.S. Senate Appropriations Committee's 1970 hearings on the following year's Bonneville appropriation. Testimony about net billing apparently won approval from committee chair Allen Ellender (D–Louisiana), but Bonneville managers provided only a cursory description of the process.[77] Thus, the financing mechanism for the first WPPSS nuclear plant (and for two more plants later on) never received formal legislative sanction – and very little public scrutiny. What seemed to Northwest energy planners in the 1960s like a fair bargain for all looked to critics in the 1980s and beyond like an evasion of Bonneville's mandated role and an invitation to irresponsibility.

In planning the three-party contracts (each involving a utility, Bonneville, and the Supply System) that net billing required, energy officials wanted to parcel out authority carefully. Since Bonneville was taking on the net-billed plant as a regional commitment, it needed the power to oversee work on the investment it was underwriting. Yet, at the same time, WPPSS needed the ability to do its job without undue interference. Similarly, participants in the Supply System's nuclear project needed influence to safeguard their investments, but WPPSS itself had to be capable of acting without the hindrance of squabbles among scores of public power utilities. The project agreements which Bonneville, WPPSS, and the utilities

---

[76] Tollefson, *BPA and the Struggle*, 353. "Hydro-Thermal Program for Pacific Northwest Approved," *Northwest Public Power Bulletin*, 23, 11 (November 1969): 3.

[77] Department of Interior, Bonneville Power Administration, *The Role of the Bonneville Power Administration in the Pacific Northwest Power Supply System Including Its Participation in the Hydro-Thermal Power Program: Appendix A, BPA Power Resources, Acquisitions, Planning and Operations* (Portland: Bonneville Power Administration, 1977), I-20–22 summarizes Bonneville's legal justification of net billing. See Michael C. Blumm, "The Northwest's Hydroelectric Heritage: Prologue to the Pacific Northwest Electric Power Planning and Conservation Act," *Washington Law Review*, 58, 2 (April 1983): 223. Jeffrey P. Foote, Alan S. Larsen and Rodney S. Maddox, "Bonneville Power Administration: Northwest Power Broker," *Environmental Law*, 6 (Spring 1976): 831–858 criticizes the legal basis of the Hydro-Thermal Program on p. 847. Net billing appears in the discussion of Bonneville's appropriations for fiscal year 1971: Senate Committee on Appropriations, *Public Works for Water, Pollution Control and Power Development and Atomic Energy Commission Appropriations*, Part 3 Volume 1. (Washington, DC: U.S. Government Printing Office, 1971), 22 and 39–41.

eventually devised appeared to please all parties. Bonneville could oversee the Supply System's work and had the right to review and approve or disapprove of major policy decisions. The utilities got a Participants' Review Board with promises of access to key information and the authority to approve budgets.

After its announcement that it would build a nuclear plant, the Supply System received a set of lessons in the complex bureaucratic politics of nuclear energy. Initially, however, planners expected a smooth path. A timetable for the necessary tasks indicated that site selection should be completed by December 1969. Interim financing arrangements and arrangements for sale of long-term bonds would then follow. Construction, according to plan, would begin sometime between September 1971 and September 1972.[78] Finding the right location was among the trickiest tasks. Siting the plant on the Hanford Nuclear Reservation would be a comfortable solution for the Tri-Cities–based Supply System, but a plant west of the Cascades, nearer the region's population centers, offered technical and political advantages. A west side project would cut losses in transmitting electricity to its users. Moreover, it might help secure cooperation between WPPSS and the municipal systems in Seattle and Tacoma that had customarily stood apart from public utility politics. By late 1969, Roosevelt Beach, on the coast in Grays Harbor County, south of Puget Sound, had emerged as the Supply System's preferred location.

However, Roosevelt Beach encountered stiff opposition. The project was to employ a "once-through" cooling procedure that would discharge warm water into the ocean. The effect of thermal pollution on aquatic life had become one of the leading issues for an anti-nuclear movement taking shape around the nation.[79] To gain a site certificate, the plant needed approval from the Washington Thermal Power Plant Site Evaluation Council (TPPSEC), a newly established state agency. In October 1970, as Owen Hurd, H. R. Richmond, and Ken Billington were discussing the WPPSS site application at BPA headquarters in Portland, Hurd received a phone call from a staff member of the Thermal Plant Council. He reported to Hurd that the State Fisheries Department objected to

[78] Owen Hurd, memo to WPPSS Board of Directors and Managers, March 13, 1969, in Box 8, Washington PUD Association Collection.

[79] J. Samuel Walker, "Nuclear Power and the Environment: The Atomic Energy Commission and Thermal Pollution," originally published in *Technology and Culture*, 30, 4 (October 1989), reprinted in Marcel LaFollette and Jeffrey K. Stine, eds., *Technology and Choice: Readings from Technology and Culture* (Chicago: University of Chicago Press, 1991), 203–231.

the Roosevelt Beach plan and would not support certification without a six-year study of the impact of warm water discharges on the area's clam population. Almost immediately, the three decided that the project had to be relocated to Hanford, and within days the Supply System Board decided to apply for a site on the nuclear reservation. The Hanford climate was hospitable; Hurd reported "excellent cooperation" from the Atomic Energy Commission, the city of Richland, and others in the area.[80]

Technicalities of the net billing arrangements proved difficult to work out. In 1969, the Public Power Council told its members that net billing arrangements should be signed by January 1970. Contracts with participant utilities were drafted by that spring, but needed clearance from the U.S. Bureau of the Budget and a congressional appropriations committee. They were not signed until January 4, 1971. Similarly, interim financing for pre-construction activities was delayed a year while the Supply System waited to hear from the Internal Revenue Service that interest on notes and bonds for the plant would be tax-exempt.[81]

In January 1971, WPPSS filed an application with the state's Thermal Power Plant Site Evaluation Council for approval of the Hanford location; hearings were scheduled for March.[82] Although no one expressed opposition at the initial meetings, the Site Evaluation Council, though friendly, moved slowly. In the summer, problems arose with plans to use an unlined cooling pond for the water discharge from the plant's generators. This might raise the level of the water table at the site too much. The next month, new WPPSS Managing Director J. J. Stein reported that the plant would have to be built with a cooling tower instead of a pond.[83] Eventually, Governor Dan Evans signed the site authorization in May 1972.

The federal government demanded more effort. Preliminary site work not requiring a full construction permit began with groundbreaking ceremonies on August 14, 1972, but for major building the Atomic Energy Commission required a detailed permit application, a Preliminary Safety Analysis Report, and 300 copies of an Environmental Impact Statement. Planning for this began in the winter of 1971, but the license

[80] Minutes, WPPSS Executive Committee, October 16, 1970, p. 3.

[81] Memo, Alan Jones to Public Power Council participants, no date [1969], in Box 8, Washington PUD Association Collection; Harry Dutton, "To Add 1.1 Million Kilowatt Resource to NW," *Northwest Public Power Bulletin*, 25, 2 (February 1971): 4–5; Minutes, Public Power Council Executive Committee, February 3, 1971, p. 4.

[82] Minutes, WPPSS Board of Directors, August 14, 1970, September 18, 1970; Minutes, WPPSS Executive Committee, February 12, 1971.

[83] Minutes, WPPSS Executive Committee, July 23, 1971, pp. 3–4; August 13, 1971, p. 2.

application was not filed until August. It took another nineteen months of sometimes discouraging delays before the construction permit came through in March 1973. The Supply System boasted that this was the shortest review period since a Federal Court of Appeals had ruled in 1971 that building nuclear plants required Environmental Impact Statements.[84] The AEC's own counsel called the public hearing in Richland, where all public statements supported the proposal, "a unique and gratifying experience." Supply System leaders appreciated the nurturing environment around Hanford. "Where else," noted Board chairman Ed Fischer at the groundbreaking, "would we find friends like a Tri-Cities Nuclear Council, beating the drums for WPPSS? And a *Tri-City Herald* [the local newspaper]...impatient to get the action started?"[85]

Three months later, on June 26, 1973, the Supply System sold its first bond issue to raise $150 million at an interest rate of 5.65 percent. Thirteen more issues were to follow in the next nine years; all bore higher interest rates than the first issue. Although WPPSS and its public power backers experienced some frustrating moments in dealing with state and federal agencies while preparing to start work on its plant, in fact the Thermal Power Plant Site Evaluation Council and the Atomic Energy Commission both acted sympathetically and fairly expeditiously. Even then, the approval process was protracted. In 1967, Owen Hurd had announced that WPPSS could finish a 1,000 megawatt project by July 1972 at a cost of $123 million. Nearly a year after that promised completion date, the Supply System finally could begin construction in earnest on a plant that was now anticipated for September 1977 at a cost of $400 million.

Nuclear Plant Proliferation

Because the Hanford Generating Project had been WPPSS's first nuclear venture, this new project was named WPPSS Nuclear Plant Two (WNP-2). By the time WPPSS began WNP-2, other nuclear plants had been marked on the Northwest's planning maps as part of the Hydro-Thermal Power Program. Portland General Electric's Trojan reactor was already under construction at Rainier, Oregon. The utility expected to have the plant on

---

[84] "Supply System Activities for Month of March 1973," March 29, 1973, p. 3, in Box 7, Washington PUD Association Collection. The court decision was Calvert Cliffs Coordinating Committee v. U.S. Atomic Energy Commission, 449 F.2d 1109.

[85] Minutes, WPPSS Board of Directors, February 5, 1973, p. 4; Fischer comments in "Index to Corporate Files" notebook, Administrative Central Files, Supply System Archives.

line by 1974. Though Portland General Electric was a private company, the Eugene Water and Electric Board (EWEB) had a 30 percent ownership share in Trojan. This capability was assigned to Bonneville under net billing contracts similar to the Supply System's.

EWEB itself had jumped into the Hydro-Thermal Program with plans for its own nuclear project. In November 1968, Eugene voters had over-whelmingly approved borrowing $225 million to finance the plant. How-ever, later that winter opposition appeared. At first, farmers in the area where EWEB initially hoped to site the plant objected; soon, as EWEB reconsidered its siting choices, a band of liberal activists formed the Eugene Future Power Committee (EFPC). The Committee at first pressed the utility board to open the project to public discussion, something the business-oriented board was reluctant to do. Later in 1969, the group launched a petition drive to put an initiative measure on the city ballot calling for a moratorium on EWEB's nuclear project. They adopted the slogan, "We can wait. We should wait."[86]

The initiative went to the voters in May 1970. This time Eugene actively debated its energy future. The Future Power Committee raised questions about health and safety dangers in nuclear power but it concentrated its fire on the specifics of EWEB's proposal. It actively challenged the utility's prediction that local demand would double between 1970 and 1976 and devised its own demand projection. Years later, Keith Parks, a top EWEB manager, admitted that the Future Power Committee's forecast had been "right on the button." The Committee won a narrow victory in May, with 51.8 percent of the city's voters approving the moratorium.

The moratorium's passage took the EWEB project off the regional drawing board. Nuclear power was emerging as a key issue for a growing environmentalist movement. Eugene, a generally liberal university town, was a center of this consciousness, but its impact would spread through-out the Northwest. Even Keith Parks changed his mind about EWEB's project. During the Future Power Committee's campaign, Parks felt they were "sabotaging something that was good." In a 1986 interview, how-ever, Parks conceded, "They did a great favor for this community. They saved its butt."[87]

---

[86] For a detailed account of EWEB's nuclear project and its opponents, see Daniel Pope, "'We Can Wait. We Should Wait.' Eugene's Nuclear Power Controversy, 1968–1970," *Pacific Historical Review*, 49, 3 (August 1990): 349–373.

[87] Author's interview with Keith Parks, Eugene, Oregon, October 16, 1986.

To Northwest public power forces, Eugene's rejection of the EWEB plant came as an unpleasant surprise. The EWEB project had been slated as the first public power nuclear plant in the Hydro-Thermal Program. The moratorium, which many correctly assumed would fade into an outright cancellation, put new strains on a schedule already experiencing "considerable slippage." A week after the Eugene vote, the Supply System convened a special Board of Directors' meeting in Vancouver, Washington, with Russ Richmond and Bernard Goldhammer of Bonneville in attendance. Those present agreed to go slowly, avoiding "unilateral action," but also agreed that "the scheduling of two public 1,000 MW Nuclear Plants in Washington on a net billing basis deserves consideration."[88]

To go beyond the one project it had already undertaken, WPPSS would have to consolidate support among the state's public utilities. The Supply System's members in 1970 consisted of seventeen Washington Public Utility Districts and the city of Richland's municipal utility. Conspicuously absent were Seattle and Tacoma, the first and third largest cities in the state, and their municipal utilities. Moreover, Seattle and Tacoma, as well as some of the larger PUDs, harbored ambitions of building their own nuclear plants. Seattle and Tacoma hesitated to take participants' shares in WNP-2, hoping to save their net billing capacity for their own future projects, and the Supply System fretted about their delays through the fall of 1970.[89] WPPSS supporters worried that the large public utilities in the Puget Sound area might form a rival Joint Operating Agency and start ventures in conjunction with private utilities.

Following the Eugene vote, Ken Billington began pressing the Supply System to reorganize in order to keep Snohomish County PUD (north of Seattle) in the fold and recruit Seattle and Tacoma as new members. The System's structure was unwieldy. The Board of Directors, with one representative from each member utility, was too large to guide policy effectively and lacked ability to channel communications between member utilities and Supply System management. WPPSS needed to take measures to expand involvement of public utilities in its activities and to arrange to share electricity supply with the investor-owned companies. The Board's Executive Committee had seldom met after 1967; the entire Board met about once a month. Billington proposed that WPPSS restructure the Board to assure an influential voice for the large utilities west of

---

[88] Minutes, WPPSS Board of Directors, June 3, 1970.
[89] Minutes, WPPSS Board of Directors, October 16, 1970, p. 3.

the Cascades. With an expanded Supply System and reorganized governance, WPPSS could continue with its Nuclear Project 2 (still at that point scheduled for Roosevelt Beach) and pick up the stalled EWEB project.[90]

Owen Hurd balked at this proposal; it seemed to surrender control to the Puget Sound utilities (Seattle, Tacoma, and Snohomish). His old friend Billington, however, viewed reorganization as an urgent need. The WPPSS Board appointed a special committee to study reorganization and voted in July to invite Seattle and Tacoma to join "in furtherance of combined and joint sponsorship of thermal plants." Two weeks later, the Board agreed to form a strong Executive Committee. The Board itself would go back to a quarterly meeting schedule, with more frequent sessions for the Executive Committee. Snohomish PUD was granted a permanent seat on the Executive Committee, along with Seattle and Tacoma's utilities should they join WPPSS. To provide a say for utility professionals, at least four of the seven Executive Committee members had to be management personnel from member utilities. Notably, however, the restructuring kept policy making entirely under the control of Washington State public utility personnel. There were no outsiders on the Board or Executive Committee.

With these developments, Billington could arrange a session with Seattle Mayor Wes Uhlman to urge him to support affiliation with the Supply System.[91] The city's response was tentative. Although it eventually joined in the net billing arrangements for WNP-2, it did not become a WPPSS member until March 1971. Tacoma was even more hesitant. It did not execute a net billing agreement for the project and joined the Supply System in 1972.[92]

Partly as a side effect of the controversy over expansion and restructuring, Owen Hurd retired as Managing Director. Hurd was already over the standard retirement age, but in 1970, the Board had extended his mandatory retirement date to January 31, 1973. However, in April 1971, Ed Fischer, the Board's Chairman, read a statement announcing Hurd's resignation, effective July 1, 1971. J. J. (Jack) Stein was named as his successor with a salary set at $35,000. Although he had been a naval

90 Memo, Ken Billington to All Commissioners and Managers, June 19, 1970, in Box 24, Washington PUD Association Collection.
91 Billington to Hurd, July 7, 1970 in Box 24, Washington PUD Association Collection. Minutes, WPPSS Board of Directors, July 31, 1970, p. 6 and August 14, 1970, p. 5. [Ken Billington], "Purposeful Comments and Responses Prepared for Discussion Purposes with Mayor Wes Uhlman and Staff on Friday, September 11, 1970," in Box 24, Washington PUD Association Collection. Billington, *People, Politics & Public Power*, 302–305.
92 Minutes, WPPSS Board of Directors, December 4, 1970, p. 3; March 23, 1971, p. 5.

officer, Stein was deeply enmeshed in the Northwest's public power net-work. Manager of the Grays Harbor County PUD in western Washington for fifteen years, he was serving on the Executive Committee of WPPSS when he was selected as Managing Director. Stein had no background in nuclear projects. The appointment of this insider reinforced the provincial character of the Supply System.

In the next two years, the thought that WPPSS might build a reactor to substitute for EWEB's defeated project crystallized into policy. While the Supply System contemplated this, a crisis impelled it to consider yet another round of expansion. The Hanford N-Reactor had originally been designed to manufacture plutonium. Dual-purpose operation was to con-tinue as long as the military required the reactor's plutonium output. Then it would be switched to the single task of generating energy. Plan-ners expected this transition would occur in the late 1970s, but national politics brought matters to a head in 1971. Washington's Governor Dan Evans, a moderate Republican, had backed Nelson Rockefeller in the 1968 GOP presidential primaries. Richard Nixon, looking for spending cuts in his fiscal year 1972 budget, apparently remembered. Nixon was also aware of the growing prominence of Scoop Jackson; the N-Reactor's godfather was eyeing a presidential run in 1972.[93] On January 28, 1971, the Office of Management and Budget directed the Atomic Energy Com-mission to close down the N-Reactor immediately.

Northwestern utility leaders and politicians were aghast. In the words of the *Northwest Public Power Bulletin*, the shutdown "caused chaos in the Northwest power situation, deepened the economic recession in Washington State, and thickened the smog in Los Angeles."[94] Perhaps the reactor could be converted to power-only use, but that would be costly and could take three to five years. Meanwhile, Senator Jackson and others warned, a power deficit of 300 megawatts or more faced the Northwest the following winter if stream flows were low. If Bonneville had to curtail power to industrial users, as many as 7,500 manufacturing jobs might be lost.[95] In the next three months, the region worked feverishly to reverse the decision. A hastily formed group in the Tri-Cities named itself the Silent Majority, appropriating the Nixonian term to protest the

[93] Pendergrass, "Public Power, Politics, and Technology," 179–180.
[94] *Northwest Public Power Bulletin*, 25, 2 (February 1971): 2. This reference to Los Angeles smog implied that the N-Reactor shutdown would curtail hydropower sales on the Intertie to California and hence require more fossil-fuel generation there.
[95] William W. Prochnau, "Big Hanford cut in store," *Seattle Times*, January 28, 1971.

shutdown.[96] While public and private utility leaders quarreled about who should bear the brunt of any shortages it caused, they united to press for a reversal.

In April, the N-Reactor won a reprieve, an agreement to continue its dual-purpose operation until July 1, 1974. In return, Northwest utilities and Direct Service Industries would pay $20 million annually to the AEC for steam to generate up to four billion kilowatt-hours. As well, the Supply System and the OMB agreed that WPPSS would study whether to take over the N-Reactor and convert it to a power-only facility.[97]

The Supply System at first expected that it would assume operations of the N-Reactor at the end of the grace period. Ed Fischer described the Supply System's purpose: "to carry on with Hanford #1 [the Hanford Generating Project] – to get on with Hanford #2 [WNP-2] and to engage in such other projects which appear to be in the best interest of an adequate power supply for public power agencies in the Pacific Northwest."[98] However, by the spring of 1972, technical studies had shown that conversion to power-only operation would not work.[99] Planners and administrators worried that implementation of the Hydro-Thermal Program was lagging. That May, at a meeting of the Public Power Council's Executive Committee, BPA Administrator Richmond "stated his concern about the Pacific Northwest utilities' lack of progress in designating sponsorship and location" of thermal plants the region would need by the 1980s.[100] Richmond told public power to "take some definitive action" to get the program "back on schedule."[101]

On May 17, 1972, the Public Power Council responded to Richmond with a resolution asking the Supply System to begin work on a new reactor for the Hanford Generating Project, to be completed by 1980, and to start on another nuclear plant scheduled to come on line in 1981. The motion passed unanimously. Nine days later, without evident opposition, the Supply System's Executive Committee approved a letter to the PPC

96 William W. Prochnau, "Closing 2 reactors made sense but…," *Seattle Times*, January 31, 1971.
97 "Power Picture Brightened by N-Reactor OK," *Northwest Public Power Bulletin*, 25 (May 1971): 14. Henry G. Curtis, "General Manager's Comments," ibid., 2–3. Billington, *People, Politics & Public Power*, 309–327, offers a detailed account of negotiations.
98 Minutes, WPPSS Executive Committee, June 11, 1971.
99 Minutes, WPPSS Executive Committee, May 12, 1972, p. 1.
100 Minutes, Public Power Council Executive Committee, May 8, 1972, pp. 1–2.
101 Minutes, WPPSS Executive Committee, May 12, 1972, p. 1.

"accepting...the responsibility for immediately undertaking" the new projects.[102]

Now there were three WPPSS net-billed plants. That September, WPPSS recognized that replacing the N-Reactor at the same site was inefficient and decided to build elsewhere at Hanford. As a replacement for the N-Reactor, it became WPPSS Nuclear Project One (WNP-1). The other new venture, designated as WPPSS Nuclear Project Three (WNP-3), was to be sited west of the Cascades.

Private utilities were to have an ownership share in the expected output of WNP-3, as they had been set to have in the canceled Eugene project. In the final arrangements, public agencies owned 70 percent of the plant capability and the four Oregon and Washington investor-owned companies (Portland General Electric, Pacific Power & Light, Puget Sound Power & Light, and Washington Water Power) divided the remaining 30 percent. Only the public share would be financed through Bonneville's net billing. WNP-1, like the initial WNP-2 project, would be fully public and fully net billed.

These decisions reflected demand forecasts that the investor-owned companies would need the large majority of new regional generating capacity, especially in the latter years of the Hydro-Thermal Program. According to forecasts presented to the Supply System's Executive Committee, from 1982 to 1992, public utilities would require completion of about three additional large thermal plants, but twelve others would be needed to supply the IOUs' customers.[103] Bonneville's preference customers, the public utilities, wanted to build enough to meet their own needs. They were happy to see the private companies undertake their own projects or even to share ownership with public agencies, as in WNP-3. The public utilities did not, for the most part, want to build plants that would end up supplying the private firms' demand growth cheaply. If Bonneville took the output of public projects under net billing and sold it to IOUs at the same bargain rate it offered the publics, this would weaken the sacrosanct principle of public agency preference. On the other hand, if the public agencies completed a plant before they needed its entire capacity, they wanted to sell its excess output to the private companies until

[102] Public Power Council Annual Participants' Meeting, May 17, 1972, p. 2, in Box 7, Washington PUD Association Collection. Minutes, WPPSS Executive Committee, May 26, 1972, p. 4.
[103] Minutes, WPPSS Executive Committee, October 11, 1972, p. 1.

customers' demand grew to match supply. Thus, they looked for term sales agreements with the investor-owned firms, yet they feared longer-term commitments.

The End of Net Billing

In May 1972, when it agreed to take on Projects 1 and 3, the Supply System estimated total costs at about $1.3 billion for the three net-billed plants.[104] They would generate more electricity than Grand Coulee Dam. The Washington Public Power Supply System had ninety-one employees at the end of 1972, the year it agreed to take on two additional nuclear projects. A few weeks later, Managing Director Stein warned the Board that the sixty-eight members of the administrative staff were "stretched mighty thin to keep track of everything that is going on." The Supply System would require another twenty workers during 1973. It would also have to tackle an office space problem: "[W]e didn't expect to be involved in three large nuclear projects when we moved into our present quarters." Northwest public power interests recognized the "tremendous responsibility" that the Supply System had assumed.[105]

The Supply System had become the chosen vehicle for building public power nuclear plants for several reasons. The Joint Operating Agency law in Washington State gave it legal authority to undertake the task. The organization, located near Hanford, imbibed a pro-nuclear viewpoint from its environment. A pervasive sense of the need for urgent action to build supply led Bonneville and the Public Power Council to turn to WPPSS when an individual utility, EWEB, lacked the political clout to move ahead independently. Finally, the Supply System, as a public agency, offered a huge financial advantage over private construction. It could borrow on the tax-exempt municipal bond market. Bondholders' interest payments would be free from federal (and in most cases state) income taxes. Interest rates on municipal bonds were lower than on comparable federal or private securities to reflect that tax advantage.

However, a cloud hung over the Supply System's tax exemption for net-billed plants. The 1950s and 1960s had seen the proliferation of

---

[104] Washington Public Power Supply System, *Annual Report 1973*, p. 9. Minutes, WPPSS Executive Committee, June 21, 1972, Attachment: letter from J. J. Stein to eleven brokerages, June 10, 1972.

[105] Minutes, WPPSS Executive Committee, March 30, 1973, p. 5; Minutes, Public Power Council Membership Meeting, June 22, 1972, p. 1.

municipal revenue bonds, secured by pledges of income from the projects they financed rather than from general tax receipts. By 1970, revenue bonds totaled a third of state and local borrowings. Increasingly, too, the revenue bonds financed undertakings beyond the traditional boundaries of government activities. In particular, many states, municipalities, and other public agencies issued industrial development bonds. Often, these issues paid for facilities to be leased to private firms. The lease income then paid off the bondholders

Industrial development bonds drew fire as public subsidies of some private enterprises at the expense of rivals. In January 1969, the Treasury Department proposed rules to limit their use. The thrust of the rules was to prevent public borrowing for the use of private companies or for the federal government, which itself lacked the tax-exempt borrowing advantage. In May 1970, all U.S. senators from Oregon, Washington, Idaho, and Montana wrote to Secretary of the Treasury David Kennedy to warn him that the proposals might halt joint public-private power supply arrangements, especially plans to sell surplus energy from public plants during their early years of operations to the private firms.[106]

On August 3, 1972, after more than three years of deliberation, Treasury promulgated its municipal borrowing limitations. For issues of five million dollars or more, municipal bonds would lose their tax exemption if a quarter or more of their proceeds were used by a private business. Public utilities' bonds would remain tax exempt "if 25% or less of the utility facility's output is used by private business. Should private businesses use more than a quarter of the output, the bonds will be taxable, even if they are issued on behalf of a municipally-owned public utility." Since the 1969 Treasury proposal had set the limit at 50 percent, the final regulations were a blow to Northwestern public utilities. To sell even a quarter of their output to IOUs would risk losing their bonds' tax exemption.[107]

When the Public Power Council's leaders convened in Seattle on August 25, 1972, and discussed the Treasury edict, "The consensus was that net billing remains as a good method to finance the public agencies' share of thermal power projects."[108] Indeed, the Internal Revenue Service did allow Bonneville, the Supply System, and 104 participating public utilities

---

[106] *Weekly Bond Buyer*, June 8, 1970, 61.

[107] "Treasury Sets New Rules for Industrial Development Bonds," *Commercial and Financial Chronicle*, August 10, 1972. The Treasury Decision became Federal Tax Regulation 1.103–7. See *Federal Tax Regulations 1973*, I (St. Paul, MN.: West Publishing, 1973), 314–321.

[108] Minutes, Public Power Council Executive Committee, August 25, 1972, p. 3.

to sign net billing agreements for WNP-1 and WNP-3 in 1973, despite the ruling. However, tax-exempt bond financing and net billing, the Hydro-Thermal Power Program's core financing mechanism, would not be available thereafter.

Even without its legal difficulties, net billing was in financial trouble by 1973. The payments that participating utilities would make to the Supply System could not exceed the gross costs of their Bonneville electricity purchases. In our hypothetical example above, the ABC Utility paid $400,000 a month to the project sponsor; BPA deducted this amount from its $1,000,000 power bill to ABC for a net bill of $600,000. As net-billed project capacities increased or costs rose, BPA's billing credits to ABC could also increase, but not beyond its gross bill to ABC. Moreover, Bonneville needed some monetary revenue from its preference customers; in practice, utilities' net billing credits could not exceed about 85 percent of their power purchase costs.

As early as June 1972, the Public Power Council exhorted all preference customers to participate in WNP-1 and 3. If some opted out, the net billing capacity of the remaining utilities might fall short of the amount needed to pay for the projects.[109] A Bonneville staff estimate in early 1973 revealed that its sales to preference customers in 1981 would total about $145 million, while costs of the net-billed plants to the participants would equal about $135 million; this would already surpass the 85 percent barrier.[110] To increase net billing capacity, Bonneville would have to raise its rates. However, net billing had been designed in the first place to help avoid large rate increases.

Combined with the Treasury Department restriction on tax exemption, the exhaustion of Bonneville's net billing capacity halted the Hydro-Thermal Power Program devised less than five years earlier. To the public power leadership in the Pacific Northwest, these years had balanced accomplishments with frustrations. Despite clashes, public and private utilities had succeeded in crafting a plan for the region's anticipated load growth and finding organizations willing to meet the program's goals. That WPPSS had decided to build three net-billed nuclear plants showed that public power could maintain its importance in the Pacific Northwest's electrical supply. After years of groping for a leading role, the Washington Public Power Supply System now had one. Yet the situation in 1973 portended serious difficulties. The West Group demand forecast that February

[109] Minutes, Public Power Council Membership Meeting, June 22, 1972, p. 2.
[110] Lee et al., *Electric Power and the Future...*, 85.

predicted over 60 percent growth in energy demand during the next decade.[111] Deficits would develop by the end of the seventies. Net billing, the key mechanism for financing new public-owned thermal plants and integrating them with the hydro system, had been undermined. Yet public power remained convinced that it needed to find a way to undertake more nuclear projects without net billing. Finally, as we shall see, the problems of building the Supply System's nuclear plants on time and within budget were already turning out to be daunting, perhaps insuperable.

[111] House of Representatives, Committee on Interior and Insular Affairs, Subcommittee on Mining, Forest Management and Bonneville Power Administration, *The Bonneville Power Administration [BPA] and Washington Public Power Supply System [WPPSS]*, 98th Cong., 2d Sess., 1984, Serial No. 98–48, Part III, 154. Hereafter cited as *The BPA and WPPSS*.

3

The Second Wave – Projects 4 and 5

Electrical utility leaders in the Pacific Northwest watched skies and weather forecasts anxiously in the first months of 1973. Below-normal winter precipitation and a dry spring meant a thin mountain snowpack, diminished runoffs, and trouble for hydropower. In April, Bonneville curtailed 500 megawatts of electricity to its Direct Service Industry customers. Hydroelectric supplies continued to shrink; on August 16 utilities announced a program of voluntary cutbacks of electrical usage. Utilities and their customers performed well, reported the Pacific Northwest Utilities Conference Committee, and demand decreased by 5.6 percent. Oregon's Governor Tom McCall issued an order banning decorative, advertising, and display lighting in the state. Still, the region faced deficits and had to import expensive power from beyond the region. Relief came with autumn rains, and heavy rainfall in November ended the shortage. But planners remained concerned. "The future outlook for the Pacific Northwest is not optimistic," the report stated. "Curtailment possibilities exist for the next four years and longer if new generation is delayed. Shortage of other fuels is occasioning conversion to electric energy, especially for residential heating and industrial process heating." In Bonneville's 1973 Annual Report, Administrator Donald Hodel described the drought as an "immediate power crisis of potentially staggering dimensions."[1]

If the 1973 drought was a regional warning, international developments that year sounded louder alarms. U.S. reliance on foreign petroleum had

---

[1] Reprinted in *The BPA and WPPSS*, 86; "Water-short Northwest thirsts for power," *Northwest Public Power Bulletin* 27, 6 (June-July 1973): 6–9; "Rains rescue draught [sic]-starved reservoirs," *Northwest Public Power Bulletin* 27, 11 (December 1973): 5; "The Pacific Northwest Electric Energy Shortage of 1973," prepared for Pacific Northwest Utilities Conference Committee in the Northwest Power Pool Coordinating Group Office, April 1974. Copy in Box 24, Washington PUD Association Collection.

grown rapidly. Crude oil output in the United States peaked in 1970 and then declined; petroleum net imports nearly tripled between 1964 and 1973.[2] Electric utilities on the Atlantic Coast, which burned more oil than those in other regions, experienced brownouts in the summer of 1970. Nationalist leaders, especially in Algeria and Libya, challenged the dominance of multinational oil companies. President Nixon's wage-price freeze of August 1971, followed by rounds of price controls, coincided with increases in world energy prices. By the summer of 1973, there were spot shortages of petroleum products, and indications that supplies of gasoline and fuel oil would remain tight. In October, within about two weeks of the outbreak of the Arab-Israeli Yom Kippur War, Arab oil ministers announced production cuts. Within a few days, Saudi Arabia and the other Arab states had curtailed all shipments to the United States. Oil prices jumped; by December, Iran sold oil in the spot market for over $17 a barrel, 600 percent higher than the price before the embargo.[3] Long lines at gasoline pumps symbolized the end of an era of abundant energy and the mobility it allowed. The embargo, soaring prices, and shortages suddenly transformed popular indifference about energy issues to almost-apocalyptic pessimism. According to one public opinion expert, there were "incipient signs of panic," because people were "growing fearful that the country has run out of energy."[4]

For the electric utility industry, the energy crisis of 1973 offered vital, though contradictory, lessons. Almost a fifth of the nation's thermally generated electricity came from oil, and that share had doubled in the previous five years.[5] President Nixon, in a November speech, called on utilities to stop switching plants from coal to oil. His "Project Independence" program emphasized coal and nuclear electric generation, using domestic fuels. Since natural gas supplies also appeared to be shrinking, the trend to electrical home heating seemed likely to accelerate. The utility industry, both public and private, thus interpreted the energy crisis as a spur to electricity consumption and to nuclear generation. However pinched generating capacity might be, the petroleum crisis portended a rapid shift

[2] U.S. Department of Energy, Energy Information Agency, *Annual Energy Review 2002*, Table 5.1, Petroleum Overview 1949–2002, http://www.eia.doe.gov/emeu/aer/txt/ptb0501.html, accessed July 7, 2004.

[3] Daniel Yergin, *The Prize: The Epic Quest for Oil, Money, and Power* (paperback ed.; New York: Simon & Schuster, 1992), 615.

[4] Daniel Yankelovich, quoted in ibid., 618.

[5] Edison Electric Institute, *Statistical Year Book of the Electric Utility Industry/1981*, no. 49 (December 1982): 30, Table 21.

from other energy forms to electricity for space heating, perhaps industrial uses, and even electric vehicles. "There is no question ... that the need for electric service will continue to grow," wrote the chairman of the private firms' trade association. He cited Federal Energy Administrator John Sawhill's statement, "Electricity today provides 26 per cent of the gross energy inputs. It should provide more."[6] A manager for Westinghouse Electric, addressing the American Public Power Association, observed, "[T]he way to solve our energy shortage will be to use more electricity in the years ahead – not less."[7] On the need for more electricity, there was no conflict between private and public power.

The anticipated shift to electricity indicated a heightened role for nuclear power. A Westinghouse ad in a trade publication summed up the industry consensus, with an element of corporate self-interest: "A national commitment to nuclear energy is required to support America's shift to an electric economy."[8] Industry spokespeople endorsed measures to hasten supply growth – simplified nuclear plant licensing, regulatory relief, less demanding construction standards, and relaxed clean air regulation for fossil-fuel plants. The Nixon and Ford administrations generally shared these positions.[9]

Yet if the embargo proved to utilities and manufacturers the need for more electricity and especially more nuclear construction, others concluded that conservation and renewable energy were the real needs. The popularity of E.F. Schumacher's tract *Small Is Beautiful* (1973) showed the appeal of what some called the "conservation ethic." Amory Lovins' 1976 article in *Foreign Affairs* and his book the next year on *Soft Energy Paths* demonstrated that one could preach conservation without pillorying comfort. Lovins' emphasis on the links between energy efficiency and international peace also demonstrated that tough-minded thinking about security could point industrialized nations down the soft path.[10] Even for skeptics, the money-saving potential of conservation and renewable energy sources was intriguing.

[6] Alvin W. Vogtle, Jr., "Electricity – A Vital Resource," *Public Utilities Fortnightly*, June 6, 1974, 28.
[7] Philip N. Ross, "Nuclear-Electric Economy: Answer to Energy Crisis," *Public Power* 32, 3 (May–June 1974): 12.
[8] *Public Utilities Fortnightly*, November 8, 1973, 12–13.
[9] See, e.g., "Nixon Calls for Plant Siting and Other Measures," *Public Utilities Fortnightly*, February 14, 1974, 27.
[10] Amory B. Lovins, "Energy Strategy: The Road Not Taken?", *Foreign Affairs* 55, 1 (October 1976): 65–96, and *Soft Energy Paths: Toward a Durable Peace*, Harper Colophon Books (New York: Harper and Row, 1977).

During and after the deep recession of 1974–75, American worries about energy subsided as real gasoline prices drifted down. Oil imports began to rise again, contrary to the objectives of Project Independence. Still, the shocks of 1973 and beyond had irrevocably changed the context of energy policy making. Formerly insulated from public scrutiny and dominated by a consensus on supply growth, issues about forms and quantities of energy would now be debated publicly in an atmosphere of conflict, suspicion and anxiety.

Phase II

In the Northwest, electricity plans and policies became re-politicized from the mid-1970s onward. The ranks no longer formed predictably along the public-private divide. The old conflicts had masked agreement on energy-intensive economic growth, capital-intensive central station supply, and building demand through low and declining rates. The complicated interest-group struggles that had raged within these boundaries did not disappear in the 1970s, but their importance receded. Production versus conservation, thermal generation versus renewable resources, and energy versus environment – these were the new conflicts superimposed on the interest-group maneuvering that had characterized Northwest energy policies since the New Deal.

Legal and financial barriers to additional net billing (see chapter 2) could not have come at a less opportune time for the region's energy planners. In February 1973, the Pacific Northwest Utilities Conference Committee, the body charged with promulgating regional electric energy forecasts, predicted a 5.30 percent annual growth rate for the following five years and boosted their prediction to 5.98 percent a year later.[11] This would have doubled demand in about twelve years. Meanwhile, the Northwest was falling off the pace of new generating supplies called for in the region's Hydro-Thermal Power Program (HTPP) of 1968.

The example of the canceled Eugene Water and Electric Board nuclear plant had shown that individual public utilities would have problems undertaking nuclear projects on their own. Ray Foleen, Assistant BPA Administrator, pointed out to the agency's customers in November 1973 that the original Hydro-Thermal accord "assumed the plants will be available when scheduled." However, virtually every major power supply

---

[11] *The BPA and WPPSS*, Part III, 156.

project in the region was falling behind schedule.[12] As Bonneville and the public utilities on the Public Power Council (PPC) sought ways to build more generating capacity, they turned once again to the Supply System. In mid-1973, the Council requested that WPPSS agree to build a fourth nuclear project, to be operational by 1984. The Supply System Board's Executive Committee responded positively, with thanks to the Council for its expression of faith in WPPSS and praise for the "degree of maturity" that both the Council and the Supply System had achieved.[13] The next spring, the PPC proposed that WPPSS put yet another plant on its agenda. Plant 4 (WNP-4) was to be paired with WNP-1 on the Hanford Reservation; Plant 5 (WNP-5) would be twinned with WNP-3 at Satsop, near the Pacific Coast. Building twin plants on shared sites promised to save nearly $400 million on construction, but the first estimates of the two new projects' costs came to $2.2 billion.[14] Contemplating such a commitment provoked questions. At one point in discussions of the fifth plant, the manager of the region's largest public utility, Seattle City Light, reported to the WPPSS Board that attorneys were worried about the financial and political hazards of starting two plants rather than one. However, the organization's sense of mission overcame reservations. Jack Stein, Managing Director, replied that it was "obvious" that there was more risk with two plants but that "there was no way to avoid" it.[15] When the Supply System's Board assembled in May 1974 to enact a resolution calling for WPPSS to construct Projects 4 and 5, there was still some doubt. One member "asked if it was really possible to have these plants in operation by 1982" (as the motion specified). Stein answered that the pairing of Plant 4 with Plant 1 and of Plant 5 with Plant 3 would enable the projects to come on line that quickly.[16]

However, without the protections that net billing had provided, Northwest utilities were not about to rush to invest in these large, expensive

---

[12] "Tomorrow's Crisis Is Today," *Northwest Public Power Bulletin* 27, 11 (December, 1973): 3.

[13] Minutes, WPPSS Special Executive Committee Meeting, Wenatchee, WA, 8 June 1973, p. 4.

[14] Minutes, Public Power Council Executive Committee, 11 April 1974, p. 2 in Folder, "WPPSS Litigation Documents from PPC, PPC Executive Comm and Misc. Minutes, 3961-1," in possession of attorney Martha Walters, lent to author. "PPC Calls for Two Tandem Nuclear Plants," *Northwest Public Power Bulletin* 28, 5 (May 1974): 5; U.S. Securities and Exchange Commission, Division of Enforcement, *Staff Report on the Investigation in the Matter of Transactions in Washington Public Power Supply System Securities* (September 1988), 46. Hereafter cited as SEC, *Staff Report.*

[15] Minutes, WPPSS Special Executive Committee Meeting, 22 February 1974, p. 3.

[16] Minutes, WPPSS Special Board Meeting, 10 May, 1974, p. 3.

projects. Congressional legislation looked like one way to regain net billing's advantages. A properly drafted act might legally authorize plants four and five to be net billed and issue tax-exempt bonds, but it could not deal with the more fundamental problem that the anticipated costs of power from net-billed plants would outstrip BPA's present power sales revenue. More drastic measures, such as permitting Bonneville to purchase electricity directly from non-federal sources, might be required. This arrangement could give utilities considering building plants the promise that Bonneville would buy their electricity and meld it with federal hydropower as BPA had agreed to do under net billing. In June 1973, a lawyers' committee of the Pacific Northwest Utilities Conference Committee (PNUCC) proposed alternatives designed to enable the Hydro-Thermal Program to continue. However, WPPSS and the public utilities were dissatisfied, fearing that the PNUCC plan would dilute public preference. Public utility attorney Jack Cluck also asserted that getting a fourth WPPSS plant off the ground was too urgent to depend on "enactment of legislation, which might not be passed at all, and if passed, might be too late. It is our present view that the project may be undertaken without passage of further legislation."[17]

By the fall of 1973, doubts about congressional action had doomed the legislative strategy. Bonneville was now shepherding Northwest utilities into a new round of power planning that could go forth without congressional action. While the legal and administrative rules had changed since WPPSS had undertaken the net-billed plants, neither Bonneville nor the utilities had reappraised the region's energy future. They continued to sound alarms about looming electricity shortages. Bonneville wanted to accomplish the Hydro-Thermal Program's original goals by revised means.

Leading Bonneville's response to the failure of the legislative approach was Bernard Goldhammer, the agency's Power Manager, who had worked out the net billing scheme. A BPA veteran since 1943, an astute student of legal and administrative detail and a shrewd practitioner of Northwest energy politics, he had respect around the region.[18] Goldhammer cobbled together an ingenious set of policies designed to provide the region with the energy he believed it would need a decade later. With three decades'

[17] Jack R. Cluck to Ed Fischer, 3 July 1973, Exhibit 158/30, in notebook of attorney Martha Walters, lent to author.
[18] Billington, *People, Politics & Public Power*, 327; Lee et al., *Electric Power and the Future*, 152. Shortly before Goldhammer's death in 1977, the Pacific Northwest Utilities Coordinating Conference cited him as the region's "Architect of Tomorrow." Other plaudits are found in an obituary, *Portland Oregonian*, 25 October 1977.

hindsight, Goldhammer's design for Phase II of the Hydro-Thermal Program seems irrelevant at best. However, at the time, Northwest energy interests hoped it would save the Program from collapse and the region from disaster.

Goldhammer outlined the principles of Phase II at a meeting of Bonneville's customers – public utilities and Direct Service Industrial firms – on November 29, 1973. As before, the federal government would continue to run its hydroelectric plants, and Bonneville would operate the region's transmission system. Others would continue to build thermal generating plants, coordinating these projects with regional plans. The BPA would take responsibility for expanding transmission lines, providing hydropower for peak demand periods, and integrating the new thermal facilities into the Northwest's power grid. But without net billing, Bonneville required some other device to knit the actions of individual utilities and agencies into a coordinated regional plan.

About two weeks after the customer meeting, BPA again gathered utilities and industrial customers in Seattle to reach agreement on a program. The December 14 "Treaty of Seattle" initiated Phase II of the Hydro-Thermal Power Program. The demise of net billing meant that Bonneville could no longer promise to merge thermal and hydropower costs in its rates or to provide a financial guarantee to investors that a failed project's costs would be recouped through BPA rates, but the pact would allow it to act as "trust agent" for the utilities. In this role, it would represent preference customers in their dealings with organizations that built and ran thermal power plants. BPA would be able to oversee the development of additional nuclear and coal plants. As Bonneville described it a few years later:

a utility would "hire" BPA to act as an agent in varying capacities, depending upon the needs of the specific utility. Services rendered by BPA could include: assistance in forecasting loads, negotiating for short- and long-term firm energy purchases, negotiating sales of energy for a utility having a surplus, preparing thermal plant operation plans, transmission from thermal plants to load centers, and scheduling of resources, storage, and thermal plant reserves.[19]

---

[19] Department of Interior, Bonneville Power Administration, *Draft Environmental Statement: The Role of the Bonneville Power Administration in the Pacific Northwest Power Supply System, Including Its Participation in the Hydro-Thermal Power Program: A Program Environmental Statement and Planning Report*, Part 1: The Regional Electric Power Supply System, (DES-77-21), July 22, 1977, II-14. (Hereafter cited as *Draft Role EIS*.)

Under these arrangements, planners hoped, Bonneville could coordinate power planning on a regional basis while public utilities could join together to finance new projects with tax-exempt bonds.

Another essential element of the plan involved bolstering regional electricity reserves. Direct Service Industries (DSIs) would provide these reserves through a new set of contracts with the BPA. Bonneville could shut off up to three-quarters of their loads to meet its other power commitments under the most dire conditions. In exchange, the firms would get an attractive new Industrial Firm Power rate for a twenty-year period, through the mid-1990s. Since most of their previous contracts were scheduled to expire by 1986, the DSIs would receive an extra decade of low-cost energy in exchange for assuming the risk of power curtailments.

Along with these procedural changes, Phase II set forth a new schedule for completion of thermal generating projects. The original Hydro-Thermal Program had established a ten-year timetable, but its delays and disappointments had made utilities hesitant to endorse a decade-long extension under Phase II. Thus, the new program was designed to bring resources on line to meet energy demands for five additional years, from 1982 to 1987. In addition to the five Supply System plants, the schedule envisioned six coal-fired generators coming on line between 1976 and 1980, all under private sponsorship. Investor-owned utilities were also to complete five nuclear reactors – the Trojan plant already underway at Rainier, Oregon; two reactors at Pebble Springs in eastern Oregon; and two at Skagit in northwestern Washington. The list of plants confirmed the regional utility consensus that the Pacific Northwest faced a supply crisis, and that coal and, especially, nuclear power would save the region from shortages and brownouts.[20]

But if Bonneville and its customers remained committed to their diagnosis and prescription, by the mid-1970s others in the Northwest were approaching the energy problem differently. Several nodes of dissent were emerging. The Eugene Future Power Committee, which had imposed a moratorium on that city's nuclear schemes, stayed on after the 1970 initiative, working to elect conservation-oriented candidates to the EWEB Board. In Seattle, community activists began to press Seattle City Light to shun nuclear projects and seek "soft path" solutions. Northwest politicians such as Oregon Fourth District U.S. Representative James B. Weaver and Washington State Senator King Lysen championed anti-nuclear causes

[20] Billington, *People, Politics & Public Power*, 348; *Draft Role EIS*, Part 1, II-15.

and questioned the Supply System's proliferating projects. The doubters' efforts helped, indirectly, to stymie Phase II.

One element of Phase II did get enacted according to plan. Since Bonneville was to invest in expanded transmission lines, hydroelectric peaking generators, and other facilities, it sought to become self-financing. Rather than rely, as it had since its inception, on congressional appropriations, in the 1974 Federal Columbia River Transmission System Act Bonneville received authority to finance its capital investments through bond sales. The measure also permitted Bonneville to purchase power during brief periods of shortage. This provision was to complement the new industrial sales contracts in giving Bonneville reserve power to meet demand growth and deal with plant outages.[21] Self-finance in the long run made Bonneville an even greater force in the region.

Project Finance without a Net

Between late 1973 and mid-1976, utility attorneys and regional energy planners sought, sometimes frantically, to devise a workable scheme to finance the two new Supply System projects in the absence of net billing. Fearing that delays would mean escalating project costs and brownouts in the early eighties, they ended up with a series of Participants' Agreements whose terms would eventually cause a monumental legal and financial tangle.

Initial funding to plan Projects 4 and 5 came in the form of short-term revenue notes which the Supply System issued in March 1974 (for $2.5 million) and August 1974 (for $15 million). Supply System utilities (the eighteen PUDs and three municipal utilities that were WPPSS members and had representatives on its board) backed the notes. But these funds only provided pre-construction start-up money. The region's other public utilities needed to be brought in so that the projects could actually get underway. Here the problems became thorny.

Despite the Phase II plans, Northwest utilities worried that acceptable arrangements to spread the risk and cost of Projects 4 and 5 might fall through. They were "reluctant (if not unwilling)" to put up funds to build large nuclear plants without some kind of "collateral contractual arrangement that at least outlines the role of BPA in the disposition of that

---

[21] Michael C. Blumm, "The Northwest's Hydroelectric Heritage: Prologue to the Pacific Northwest Electric Power and Conservation Act," *Washington Law Review* 58, 2 (April 1983): 226. Lee et al., 120–121.

output."[22] Could this be devised "in time to permit the orderly financing of these project [WNP-4 and 5] on the accelerated basis requested of the Supply System?"[23] Rather than risk future energy shortages due to financing delays, regional public utilities (working mainly through the Public Power Council), WPPSS, and Bonneville started working to devise direct, two-way contracts between the utilities and the Supply System for "slices of the pie" from the new nuclear plants. As 1974 progressed utilities faced the unpalatable alternatives of postponing the projects until Bonneville's trust agent role could be arranged or going ahead with what they viewed as necessary undertakings without BPA's safety net.

Smaller utilities had special reason to look askance at the new projects. Their contracts with Bonneville guaranteed them their full requirements for power up to twenty-five megawatts. Even if BPA started curtailing power deliveries around the Northwest, these "requirements" customers seemed to be protected against cutbacks. Why then should they risk buying into WNP-4 and 5? As one official of a small utility told the Supply System, "I have a contract now that says you are going to supply power requirements until you give notice." Thus, "at no time did we say we would sign it [a Participants' Agreement] and forfeit what we have."[24] In late 1974, the Public Power Council's attorney commented that he now expected only about thirty-five to forty municipalities and PUDs to take part in the new plants, out of more than one hundred who were eligible.[25] In all likelihood, this would not be enough to finance the projects.

By the end of 1974, merely to continue preliminary work on the new projects, the Supply System needed an infusion of money beyond the sums it could raise from its own members. Promoters decided to draw up Option Agreements, which would give signers the option to purchase a share of a rather indeterminate commodity, "project capability." Those agreements would in turn serve as security for WPPSS bonds for $100 million in development funds. Lawyers drafted the Option Agreements with "take-or-pay" clauses. Whether or not the Supply System was able to complete the projects, the utilities signing the agreements would be obliged to pay their shares of debt service on the bond issue. WPPSS distributed

---

[22] Norman A. Stoll to R. Ken Dyar, 13 June 1974, Exhibit 160/105 in Notebook of attorney Martha L. Walters, lent to author.

[23] Stoll to Dyar, 13 June 1974, Exhibit 163/25 in Notebook of attorney Martha L. Walters, lent to author.

[24] Minutes, WPPSS Special Executive Committee Meeting, September 26, 1974, p. 6.

[25] Norman A. Stoll to R. Ken Dyar, December 5, 1974, Exhibit 174/119 in Notebook of attorney Martha Walters, lent to author.

them to utilities in the spring, and ninety-three of them returned signed contracts, all dated July 22, 1975. Two days later, the Supply System issued the development bonds, with Merrill Lynch as lead underwriter, at an interest rate of 7.04 percent. They received investment-grade ratings from Standard & Poor's and Moody's, the nation's major bond-rating agencies. That ninety-three public utilities in the Northwest had chosen to take options encouraged nuclear power advocates.

Although the Option Agreements took some of the heat off the Supply System, they nevertheless indicated just how pressed the project backers felt. As Bonneville sent the Option Agreements to customers, it warned them that it was "imperative" to execute and return them by May 15 (a deadline that they missed by over two months).[26] "Time is of the essence," noted the Agreement's preamble.[27] The pacts pledged the Supply System to have final participation agreements in hand by August 1, 1976. Meanwhile, they continued to hold out the hope that Bonneville would be able to furnish the services contemplated under the Phase II plan. Execution of Participants' Agreements would take place, the Option contracts noted, after arrangements with Bonneville were worked out.[28]

However, a month after the Option Agreements were returned, a complex federal District Court decision scuttled Phase II and further complicated plans to launch Projects 4 and 5. The Achilles heel of Phase II proved to be the provision for new industrial power sales contracts. Unless Bonneville could curtail its supply to the Direct Service Industrial firms, it could not satisfy its preference customers' electricity demands in times of stringency. Unless it could meet those demands, utilities would have little incentive to have Bonneville coordinate their power planning and represent them in dealing with generating facilities. Thus, the Industrial Firm contracts undergirded the agency provisions of the Treaty of Seattle. However, when the court held up plans for construction of a new aluminum plant, a series of political and legal dominoes began to topple, blocking the planned industrial sales agreements. By 1976, they had stopped Phase II in its tracks.

A decade before the rise and fall of Phase II, Bonneville had anticipated surplus electricity. In 1966, it signed a twenty-year contract to supply

[26] Bonneville Power Administration, memo to Public Agencies, April 3, 1975, Exhibit 127/72 in notebook of attorney Martha Walters, lent to author.
[27] SEC, *Staff Report*, 325.
[28] Option and Services Agreement, City of Drain, March 21, 1975 (but signature date on p. 30 is July 22, 1975), Exhibit 129/1 in Notebooks of attorney Martha Walters, lent to author.

a consortium that planned to build an aluminum plant in the Northwest. The consortium moved slowly and underwent reorganization. Their intention to build in Warrenton, Oregon, near Astoria at the mouth of the Columbia fell through when environmental restrictions and tightened limits on tax-exempt bond financing made the site unworkable. In 1974, the company, renamed Alumax, sought to relocate its plant to Umatilla in north-central Oregon.

The move east angered business interests in the Astoria region who wanted the benefits a new aluminum smelter might bring and failed to placate environmentalists, who objected to Bonneville's contract to provide electricity to this new industrial customer. When BPA revised its contract with Alumax to reflect the shift to Umatilla, it required the company to accept the new terms of Phase II direct sales contracts. This gave both sets of foes – the Natural Resources Defense Council, representing an environmentalist coalition, and the Port of Astoria, speaking for area businesses – an entering wedge. They claimed that the new power delivery arrangements constituted a federal action that would substantially affect the environment. Hence, they reasoned, Bonneville would have to prepare an Environmental Impact Statement (EIS) under the National Environmental Policy Act (NEPA). Ruling on the business groups' suit in August 1975, Federal District Judge Otto Skopil agreed, stating: "BPA's delivery of power permits the operation of this plant and federalizes this essentially private project. This type of federal involvement has consistently been held subject to NEPA."[29] The District Court indicated that an EIS dealing with Alumax alone would not suffice. Nor would it be enough to assess the industrial sales contract policy. Bonneville would have to review its entire role in the region's electric power supply system. Meanwhile, legal and administrative arrangements to implement Phase II would have to be suspended. In early 1976, BPA Administrator Donald Hodel conceded in a press release that Phase II was in "shambles."[30]

Preparing the Role EIS was a major task for Bonneville. Its draft, consisting of two main volumes, three hefty appendixes and a separate volume devoted to Alumax, was not completed until July 1977. The final version took an additional three years to appear.[31] In a changing environment,

[29] Port of Astoria v. Hodel, quoted in Jeffrey P. Foote, Alan S. Larsen and Rodney S. Maddox, "Bonneville Power Administration: Northwest Power Broker," *Environmental Law* 6 (Spring 1976): 848.

[30] Ibid., 844.

[31] United States Department of Energy, *Final Environmental Impact Statement: The Role of the Bonneville Power Administration in the Pacific Northwest Power Supply System*

where rapid growth slowed to near stagnation and projected power shortages transmuted into a regional surplus, politicians and energy interests
looked for new ways to coordinate the region's electric power planning.
In the early 1970s, Phase II had been designed to avoid having to go to
Congress to ask for new authority for Bonneville. Later, we shall explore
the complex jockeying that eventually led to the Pacific Northwest Electrical Power Planning and Conservation Act of 1980.

For utilities contemplating Projects 4 and 5, the immediate implication
of Judge Skopil's decision was that they could not count on Bonneville
to underwrite the risks of these ventures. Following the Alumax ruling,
the Supply System, Bonneville and the Public Power Council continued to
look for ways to convert the Option Agreements into workable arrangements for getting the new projects underway. Three days after the decision,
Ken Dyar, General Manager of the Public Power Council, drafted a statement calling for legislation to give Bonneville authority to purchase power
through long-term contracts. With that, BPA would not have needed the
agency agreements proposed in Phase II and could have guaranteed a
market for the electricity from Plants 4 and 5.[32] This suggestion, reversing the public utilities' long-standing aversion to allowing Bonneville to
buy energy from non-Federal sources, went nowhere but indicated the
depth of public power's anxiety about getting the projects moving.

Bernard Goldhammer, by 1975 a private consultant to Bonneville's
industrial customers, returned to the fray after Alumax with another proposal to restore BPA to the center of regional electricity development. His
"Round Three" plan, presented in a memo to Hydro-Thermal Power Program members in November 1975, would have required legislation giving
Bonneville the right to purchase non-federally generated power, including,
in particular, output from the five WPPSS plants already in the works plus
a sixth Supply System project. Skeptical environmentalist lawyers argued
that Round Three would make Bonneville the "undisputed power broker
for the region."[33] But the varied regional power interests had neither the
time nor the underlying consensus needed to put Bonneville back into

Including Its Participation in a Hydro-Thermal Power Program (December 1980),
DOE/EIS-0666. Hereafter cited as *Final Role EIS.*

[32] R. Ken Dyar, "Draft," Exhibit 131/76 in Notebooks of attorney Martha Walters, lent to
author.

[33] Memo, Bernard Goldhammer to Participants of the Hydro-Thermal Power Program of
the Pacific Northwest, 24 November, 1975, p. 17, in Exhibit 135/80 of Notebook of
attorney Martha Walters, lent to author. Foote et al., "Bonneville Power Administration:
Northwest Power Broker," 854.

the driver's seat, and Goldhammer's scheme never progressed beyond the proposal stage.

The months following the Alumax decision were anxious times for the Supply System. The $100 million raised through the Option Agreements of July 1975 would soon be spent. Those agreements looked toward signing final Participants' Agreements within a year, but would there be enough time to draw up arrangements acceptable to the option holders? To keep the projects moving forward in the interim, the Supply System proposed a "Second Option" agreement to raise another hundred million. Without this, the System's work "must abruptly cut back beginning at the end of June – with the attendant risk of very substantial money losses and of foreseeable deficiencies in the region's power supplies."[34] Urgency sometimes bordered on desperation. In a letter from Jack Cluck to Jack Stein, the attorney wrote, "[U]nless drastic steps are introduced to improve our procedure, we are likely to run into disaster." Further setbacks to the WNP-4 and 5 schedule would "require that they be scrubbed and recognized as dry holes."[35]

WPPSS 4 and 5: Seduction?

The purpose of the Treaty of Seattle and Phase II was to get more power plants built. Bonneville's preference customers had invested in the net-billed plants with little hesitation, because the financial arrangements provided an implicit federal guarantee of the debt. However, without net billing, deciding whether to participate in additional nuclear power projects was a weightier choice. Between 1973 and 1976, in a setting marked by international energy crisis, stagflation, and an international movement against nuclear power, Bonneville, the Public Power Council, and WPPSS itself worked feverishly to get local utilities to take part in Projects 4 and 5. In the end, a few utilities balked, but eighty-eight of them did sign Participants' Agreements in 1976. In 1982, those projects were terminated and the participants faced the unappealing prospect of repaying $2.25 billion in WPPSS municipal bonds for energy they would never receive. Some of the utilities sued the Bonneville Power Administration in 1983, asserting that BPA had improperly seduced them into buying

---

[34] Stoll to Dyar, March 5, 1976. Exhibit 169/41 in files of attorney Martha Walters, lent to author.

[35] Cluck to Stein, January 9, 1976. Exhibit 168/40 in files of attorney Martha Walters, lent to author.

shares of the plants. Although the suit lost in court, we can scrutinize the evidence historically to judge the forces which led the Northwest's utilities into the legal and financial morass of WNP-4 and WNP-5.

Significantly, the drive to arrange participation in projects four and five depended on a judgment that, without the new construction, the Northwest would be unable to meet energy needs in the 1980s. Demand forecasts helped create a feeling of urgency.[36] Bonneville indirectly controlled regional energy forecasts working through the Pacific Northwest Utilities Conference Committee (see chapter 2), and the forecasts in turn sought control, not just prediction, of energy futures. In retrospect, the PNUCC forecasts grossly overestimated demand.

A decade after the Supply System embarked on Projects 4 and 5, and following their termination and the System's 1983 default, bond-holding plaintiffs employed the distinguished economists Carl Kaysen and Franklin M. Fisher to study whether participating utilities' failure to use advanced forecasting techniques had resulted in their not revealing "substantial uncertainty" about the viability of the projects. A 1992 article by Fisher and several other economists outlined the approach, one that took into account price elasticity of demand for electricity and the likelihood of cost overruns on nuclear construction. The authors concluded that, in most scenarios based on the data available in the mid-1970s, participants would find themselves needing rates of over 17 cents per kilowatt-hour to cover their share of project costs. Since that price was far higher than any actual electric rate in the country, the authors concluded that the advocates of the new projects had not exercised due diligence and had failed to divulge the riskiness of their venture. Disclosure of the risk "would have meant that the bonds would not have been issued and that WPPSS 4 and 5 would not have been started when they were. They should not have been."[37]

The economists' judgment is severe but it is carefully argued and documented. It is, nevertheless, a retrospective evaluation. It does not resolve the legal question of the participants' liability nor does it consider why the region's energy planners employed the flawed forecasting methods.

---

[36] For a more complete discussion of forecasting, see Daniel Pope, "Demand Forecasts and Electrical Energy Politics: The Pacific Northwest," *Business and Economic History* 22, 1 (Fall 1993): 234–243.

[37] Franklin M. Fisher, Peter S. Fox-Penner, Joen E. Greenwood, William G. Moss and Almarin Phillips, "Due Diligence and the Demand for Electricity: A Cautionary Tale," *Review of Industrial Organization* 7 (1992): 117–149. The quote is on p. 142.

However faulty these procedures were, they did have one strong selling point; until about 1973, they had worked. Electric load growth had marched forward at a steady pace, seemingly in lockstep with advancing real output of the Pacific Northwest economy, and at rates very close to the PNUCC forecasts. Loads tripled between 1955 and 1975. Electric home heating and the growth of the Northwest aluminum industry led the way.

Northwest power planners saw this history as destiny, and they pronounced it good. They took past correlation of electricity usage and output to mean necessary causation. In one publication, Bonneville reached back to prehistory for a chart presenting the "Growing Energy Consumption from Primitive to Technological Man." Another figure plotted gross national product versus per capita energy usage for fourteen nations. That U.S. energy consumption levels were almost twice as high as Sweden's while income levels were nearly the same went unmentioned.[38] Bonneville called predicted demand levels "requirements," assuming there were no alternative to building all the generating facilities to meet them. The language is telling. Planners imbued their forecasts with a rhetoric of inevitability. The forecasts themselves "required" a supply-oriented strategy.

Demand forecasts were at the heart of BPA's pressure for new plant construction. It continually maintained that additional thermal facilities were the region's only hope. In 1972, even before the demise of net billing, Don Hodel (then the Deputy Administrator) warned preference customers of supply problems and then added ominously:

You may be thinking that "we don't have to worry about it. We are protected by the preference clause and we will be able to go on buying power from BPA...."
Consider this: In a time of regional shortage the preference clause may not mean very much. For then it may come to a political decision as to who gets the power.[39]

---

[38] Department of Interior, Bonneville Power Administration, "The Electric Energy Picture in the Pacific Northwest," May 1976, reprinted in Howard Gordon and Roy Meador, eds., *Perspectives on the Energy Crisis*, 2 (Ann Arbor, MI: Ann Arbor Scientific, 1977): 402–3.

[39] Cited in "Complaint for Damages Resulting from Breaches of Contracts and Unconstitutional Taking of Property Rights," Public Utility District No. 1 of Clallam County, Washington; and Public Utility District No.1 of Snohomish County, Washington, Plaintiffs, vs. United States of America, reprinted in House Committee on Interior and Insular Affairs, Subcommittee on Mining, Forest Management and Bonneville Power Administration, *Bonneville Power Administration: Financial Fallout from Termination of WPPSS Nuclear Projects 4 and 5*, February 14–15, 1983, 98th Cong., 1st Sess., Serial No. 98-1, p. 152. Hereafter cited as *Bonneville Financial Fallout*.

As Administrator, Don Hodel said in 1974 that the Hydro-Thermal Program was "about all we've talked about since 1969."[40] Hodel and others were predicting tight energy supplies in the 1980s even if all projects moved forward on schedule; without their success, the situation would be grave. He showed little sympathy with environmentalists who endorsed vigorous energy conservation and lashed out at them in a 1975 speech, calling them "a small, arrogant faction [of]... anti-producers and anti-achievers." He labeled them the "Prophets of Shortage," although he was the one predicting scarcity.[41*]

Hodel put strong pressure on the public utilities to buy into WPPSS Four and Five. Bonneville dangled the threat of a "Notice of Insufficiency" over their heads. Such a proclamation would give a legally required seven years' advance warning to preference customers that Bonneville would not be able to serve all their power needs after a given date. In March 1973, Interior Secretary Rogers Morton directed Bonneville to deliver this message informally. A year later, BPA again told utilities unofficially, "Bonneville will no longer furnish its preference customers' power requirements after July 1, 1983."[42]

Because a formal Notice of Insufficiency would require BPA to devise a procedure for rationing its power, the agency wished to postpone official issuance as long as possible. Yet, as one local public utility district commissioner later commented, the Notice of Insufficiency was "a gun... at the commission's head in 1975 and '76."[43] In May 1975, Hodel informed the utilities that he would delay the notification for another year. But, in the spring of 1976, as pressure for final agreement on WNP-4 and 5 financing mounted, Bonneville held a series of meetings around the Northwest outlining its intent to issue the Notice. When it finally went out on June 24, it was almost an anti-climax.

---

[40] Don Hodel, "On the Threshold – Or the Brink," *Northwest Public Power Bulletin* 28, 7 (August 1974): 7.

[41] Don Hodel, "The Prophets of Shortage," Remarks to the City Club of Portland, Portland, Oregon, July 11, 1975, reprinted in *The BPA and WPPSS*, Part III, 124.

[42] Howard Gleckman, "WPPSS: From Dream to Default," *The Bond Buyer*, reprinted in *The BPA and WPPSS*, Part I, 160.

[43] Peyton Whitely, "Snohomish County PUD to sue BPA over N-plants," *Seattle Times*, October 5, 1982.

* There were echoes of Hodel's evangelization for nuclear projects in his later career. After serving as Secretary of Energy and Secretary of the Interior in President Reagan's cabinet, he became in 1997 the President of Reverend Pat Robertson's Christian Coalition. In 2003, he was named CEO of Focus on the Family, another right-wing evangelical organization. His career thus links economic and social conservatism.

The preference customers could avoid power curtailments in the 1980s only if they took shares in the new WPPSS projects. This was the message that Bonneville continually preached. Announcing a series of meetings with regional preference utilities in fall 1974, Hodel spoke ominously:

Any utility which needs additional power resources in the mid-1980s will need to enter the Participants' Agreements with WPPSS at this time. Only by utilities sign-ing these agreements can these generating projects be constructed on the schedule required to meet the loads of Northwest utilities after July 1, 1983.

Paternally, Hodel told his customers that Bonneville had summoned them "in order that you may more fully understand your need for these agree-ments...."[44]

For some municipal utilities, PUDs and electric cooperatives, Bon-neville's campaign may have been decisive in winning acceptance of par-ticipation. In 1982, Harold Hurst, the mayor of Heyburn, Idaho, a city with fewer than a thousand ratepayers, recalled in an affidavit that he had written a letter to the Supply System in February 1974, indicating that he "was convinced that Heyburn's participation in Projects 4 and 5 was unnecessary and that the needs and requirements of Heyburn could be adequately filled by" BPA. Soon, however, BPA leaders were meeting with Heyburn officials "urg[ing] and virtually insist[ing] upon Heyburn's par-ticipation. Based on the representations of the Bonneville Power Admin-istration employees and officials, and having no other adequate sources or opportunities to evaluate the information given, the City of Heyburn changed its position." Hurst testified that it was Bonneville itself, not WPPSS, that kept the heat on, and that at least ten top BPA executives, including Administrator Hodel, had contacted the town's officials.[45] The attorney for the small Oregon coast town of Bandon recited a similar tale: "The Bandon City Council was extremely reluctant but based on the advise [sic] of BPA and under the threat that if you don't partici-pate we may not be able to supply your power needs, they finally agreed to participate."[46] Peter DeFazio, a long-time foe of the WPPSS projects and since 1986 an Oregon Congressman, claimed in a 1989 interview that "Don Hodel created this whole thing.... He went around and intimidated

[44] Hodel to BPA Preference Customers, November 8, 1974, reprinted in *The BPA and WPPSS*, Part III, p. 89.

[45] Affidavit of Harold R. Hurst, September 9, 1982, in *The BPA and WPPSS*, Part III, 400–402.

[46] Affidavit of Myron D. Spady, September 9, 1982, in *The BPA and WPPSS*, Part III, 398.

the individual utilities using ... phony [demand] projections." Hodel was, in DeFazio's words, the "mother, father, the doctor that delivered" the projects.[47]

Hodel's aggressive salesmanship cannot be discounted, but Bonneville leaders saw things differently. According to Ray Foleen, Bonneville's deputy director at the time, "Bonneville was not involved. But WPPSS and the Public Power Council had problems, they asked us if we would send out documents, talk to people. We saw the utilities every day, so we agreed. However, we weren't selling, we were acting as messenger."[48] Hodel contended that at least some of the public utilities had wanted to build nuclear plants themselves without BPA involvement. He also maintained that he had had his own trepidations about financing the two new projects. Finally, he noted that about a quarter of the eligible public utilities in the Northwest had not bought into the projects. Bonneville's enticement, he suggested, had hardly been irresistible.[49]

Indeed, BPA was not the only organization interested in rounding up local utilities to take shares of Projects 4 and 5. The Public Power Council, representing the preference customer utilities, worked alongside Bonneville to advance the projects, and the Supply System itself also sent representatives to utility meetings throughout the Northwest to encourage participation. From all quarters the message was that utilities would have to act swiftly to ensure adequate power in the 1980s.

Regardless of the pressure, the utilities had good reasons to hesitate. Under the net billing arrangements for the first three projects, if a plant could not be completed or operated properly, Bonneville would absorb its expenses into its regional rates and spread the costs to all its customers. But what would happen to utilities that invested directly in an unsuccessful project without net billing? Would they be held responsible for their full share of the project? Supporters of the new plants assuaged potential participants' fears by pointing to the promise of regionalization in Phase II plans. In December 1973, following discussions with BPA, Ken Billington, Executive Director of the Washington Public Utility Districts Association, told his members:

It was agreed that no one utility or group of utilities should have to bear losses on a plant being built for a regional purpose which does not materialize and which results from factors beyond the control of the involved utility or utilities.[50]

47 Author's interview with Rep. Peter DeFazio, Eugene, Oregon, August 16, 1989.
48 Foleen quoted in Gleckman, "WPPSS: From Dream to Default," 162.
49 Portland *Oregonian*, May 21, 1983; Tollefson, *BPA and the Struggle*, 390.
50 "Complaint for Damages," in *Bonneville Financial Fallout*, 157

Billington was repeating such assurances in 1975, indicating that BPA's conception of Phase II included "a guarantee in case of possible default."[51] However, by the time that public utilities signed up as participants in the plants in the summer of 1976, Phase II had already been derailed by the Alumax case. Legally, it appears that Bonneville never did make a binding promise to cover or reduce participants' losses on WNP-4 and 5 if they were to fail.

Among the tasks of Bonneville and the Supply System in their campaign to win adherents to the new nuclear reactors was ensuring that the scheduling of the projects would not burden participating utilities. With Plant 4 expected to begin operations in March 1982 and Plant 5 in April 1984, they would be coming on line before many of the region's public utilities expected to need the new power. Their forecasts indicated that they would be able to continue to rely on their Bonneville supplies and other resources for several years after the plants' completion. Without agreements to sell off the anticipated surplus capacity of the plants until preference customer demand grew to absorb it all, costs would be untenably high. The publics balked at committing themselves to the new plants until there was a solid plan to dispose of the excess power.

Bonneville and WPPSS offered to arrange for the industrial power users of the Northwest to take the excess output of Plants 4 and 5, potentially for the first decade of their operations. Initially, Bonneville hoped to broker a sales agreement between the participating public utilities and the industrial power users through its role as "trust agent" under Phase II. When Phase II was blocked, the Supply System itself became the intermediary. The legal logistics were complex. In the end, the participating utilities assigned their predicted surplus project capability to the Supply System, and WPPSS in turn negotiated Short Term Sales Agreements with the industrial companies.[52] Meanwhile, WPPSS and Bonneville also held out the hope that temporary surpluses could be sold south. WPPSS reported to the preference customers that the State of California and the Los Angeles Water & Power Board had expressed interest in buying electricity from the projects.[53]

The marketing of WPPSS 4 and 5 also entailed measures to diversify participation in the projects. Initially, the Supply System set out to sell a 30 percent ownership share in WNP-5 to the region's private utilities, just

[51] Ibid., 158.
[52] Examples of the Short Term Sales Agreements and the Assignment Agreements are found in Exhibits 174 and 175, Notebook of attorney Martha Walters, lent to author.
[53] These inducements are found in a question and answer sheet, "Public Agency Participation in Projects 4 & 5," reprinted in *Bonneville Financial Fallout*, 245–248

as it had done with the net-billed WNP-3. At the same time, it prepared to
package public participants' shares in Plants 4 and 5 with shares totaling
15 percent of the capability of two nuclear plants which Puget Sound
Power & Light was planning to build at Skagit, northeast of Seattle.[54]
These measures would advance integration and cooperation among public
and private power interests. They would also spread the risks of failure,
delay, or cost escalation. However, the plans hit snags. Only Pacific Power
& Light agreed to buy a 10 percent piece of Project Five, so the rest of the
project was left in the public utilities' hands. Uncertainties surrounding
Puget Power's Skagit developments kept the Supply System and Bonneville
from marketing shares of those projects with the new WPPSS contracts.[55]
Eventually, Puget was to scale back and then abandon its nuclear plans for
Skagit.[56] Unlike WPPSS 4 and 5, however, construction had never started
on these projects.*

Complicating matters further, the Public Power Council and the Supply
System itself were both signaling that WPPSS Plants 4 and 5 would not
be the last. References to future Projects 6 and 7 dot the Supply System's
records in the mid-1970s.[57] Indeed, the Supply System during the 1970s
began to consider its role to be the primary producer of energy in the

---

[54] See, for example, Minutes, WPPSS Special Executive Committee Meeting, Seattle, July
12, 1974, pp. 3–4.

[55] The Supply System announced it would not link its plans to Skagit in a memo, J. J. Stein
to All Parties to the Option and Services Agreements, April 15, 1976, Exhibit 138 in
Notebook of attorney Martha L. Walters, lent to author.

[56] On the demise of Skagit, see, for example, Bob Lane, "Skagit County Voters Say No to
Nuclear Power," *Seattle Times*, November 7, 1979; Victor F. Zonana, "Puget Sound P&L
to Cancel Nuclear Facility," *Wall Street Journal*, August 31, 1983; Daniel Jack Chasan,
"Puget Power Finally Throws in Its Nuclear Towel," *Seattle Weekly*, September 7, 1983.

[57] See, for example, Minutes, WPPSS Special Executive Committee Meeting, December 19,
1975, p. 9, which discuss a Public Power Council siting study for Plants 6 and 7.

* The Skagit projects had a history quite typical of nuclear plants planned in the mid-
seventies. They received a state site license in 1977, and hearings for construction permit were underway in 1979. However, environmental problems, along with slackening
demand and escalating nuclear construction costs, then derailed the projects. In a 1979
ballot measure, Skagit County voters voted overwhelmingly against the plants. The site
license was revoked and the projects were moved in 1980 to the Hanford Reservation,
where public support for nuclear plants remained unshaken. When the cost estimates
topped three billion dollars, Puget Power scaled back its plans to one plant, and in the
summer of 1983 abandoned this project. Puget had spent about $120 million before can-
cellation, and other regional private utilities, which had taken smaller shares, had paid
out smaller amounts. A year before, Portland General Electric had canceled plans to build
two nuclear plants at Pebble Springs in northeastern Oregon. See, e.g., Bob Lane, "Skagit
County Voters Say No to Nuclear Power," *Seattle Times*, November 7, 1979; Victor F.
Zonana, "Puget Sound P&L to Cancel Nuclear Facility," *Wall Street Journal*, August 31,
1983; Daniel Jack Chasan, "Puget Power Finally Throws in Its Nuclear Towel," *Seattle
Weekly*, September 7, 1983.

Pacific Northwest. Robert Ferguson, who became Managing Director in 1980, recalled, "They had this notion that they were going to be the energy supplier to the region, and that the [already-agreed] nuclear projects were just a part of it."[58]

Whatever the legal merits or deficiencies of the claims that Bonneville had seduced the Northwest's public utilities into participation in WNP-4 and WNP-5, several points seem clear. Bonneville did vigorously encourage participation and issued dire warnings about the consequences of failing to get these projects underway. On the other hand, the utilities were not simply passive victims of the machinations of others. The signers of the Participants' Agreements were part of a regional culture that prized economic growth and saw the expansion of electrical supply as the necessary means to achieve it. Moreover, they were part of a public utility institutional subculture that perceived the Supply System, a public agency, as the appropriate mechanism to deliver the needed energy. If they were deluded – and in retrospect they quite clearly were – the eighty-eight utilities that endorsed Participants' Agreements for WNP-4 and 5 were in part self-deluded. Yet it appears likely that Bonneville's false carrots (the hope of regionalization of the projects costs and risks) and real sticks (the threats of shortages and curtailment) pulled some utilities into the morass that these two nuclear projects became.

The Participants' Agreements

By spring 1976, the Supply System decided to prepare Participants' Agreements to offer the preference customers. With the Alumax case shunting BPA aside, legally if not politically, the agreements took the form of two-party contracts between individual utilities and WPPSS. What the Supply System offered was not electricity itself but shares of the projects' "capability." When, six years and $2.25 billion later, WPPSS terminated the projects, legal and political controversy about "capability" became the order of the day. According to the Participants' Agreements, capability was:

the amounts of electric power and energy, *if any*, which the Projects are capable of generating at any particular time (including times when either or both of the Plants are not operable or operating or the operation thereof is suspended, interrupted, interfered with, reduced or curtailed, in each case in whole or in part for any reason whatsoever), less Project station use and losses [italics added].[59]

---

[58] Author's interview with Robert Ferguson, Kennewick, WA, October 15, 1992.
[59] SEC, *Staff Report*, 278.

Participants in the 4/5 agreements purchased this capability with an obligation to "make the payments ... to Supply System under this Agreement *whether or not any of the Projects are completed, operable or operating* [italics added]."[60]

Colloquially, the clause requiring payment even for failed projects made the Participants' Agreements into "dry hole" contracts. As the name implies, dry hole provisions are associated with speculative drilling for petroleum. Given the uncertainties of oil prospecting, there is a logic to binding investors to pay their share of costs whether or not anything of value results from the project – to take the petroleum discovered or to pay for the failed venture. In the electric utility industry, take-or-pay contracts had been used rather frequently in capital construction projects. For utilities, however, they raised both legal and political issues, since building a power plant was seemingly a much less speculative project than drilling an oil well. Although some states had legislation explicitly permitting utilities to sign such agreements, Washington, Idaho, and Oregon did not. In the event of a failed project, a utility that had agreed to pay the Supply System for its share of project capability "come hell or high water" (a phrase sometimes used to describe take-or-pay contracts) would apparently have to make outlays without getting any electricity.

The take-or-pay provisions were not the only elements of the contract that left some public utilities squeamish about the legal arrangements. Since the late nineteenth century, virtually all states and municipalities have had legislatively imposed debt limits on any bonds that created general obligations against the issuers' tax receipts. Would the Supply System's borrowings become general obligations of the participants and hence potentially exceed these utilities' imposed limits? Or were they bonds secured by a specific stream of revenue (in this case, the payments of utility customers) into a "Special Fund" dedicated to meeting the bonds' obligations? The Special Fund doctrine had evolved in the twentieth century to allow the issuance of revenue bonds, where repayment was linked to income generated by the project being financed. Utility attorneys around the region studied financing plans with such questions in mind.

The Participants' Agreements also contained "step-up" provisions; if some of the participating utilities defaulted on their bond repayments, the other utilities' shares of project capability would be automatically

---

[60] SEC, *Staff Report*, 277.

increased proportionally to cover the share of the defaulting parties. This automatic step-up could have increased a participant's share by as much as 25 percent.[61] Thus, a utility which had decided to take a 4 percent share of the projects might find itself saddled with a 5 percent stake.

A related cause for anxiety was whether the Participants' Agreements between the utilities and the Supply System constituted a loan of credit. The developmental fevers that swept nineteenth-century America had inspired state and local support for projects – roads, canals, harbors, railroads – yet the historical landscape was littered with scandals, bankruptcies, and defaults, and with taxes to pay for them that citizens considered burdensome and unjust. Along with these problems, public finance since the Jacksonian era had been constrained by fears that politicians would play favorites in a competitive economy. States and municipalities often had made disastrous choices when they borrowed money for railroad promoters' and other developers' use. In the depression of 1873–79, perhaps a fifth of all municipal bonds went into default, mostly from state-aided railroad ventures that failed. Not surprisingly, state constitutions from the nineteenth century usually came to include clauses banning governmental bodies from, in effect, borrowing funds for the use of private parties. Public borrowing was to be for public purposes, not a device to finance pet special interests without public control.[62]

In the case of Projects 4 and 5, in buying "capability," were the utilities actually purchasing power through the contracts they were asked to sign, or were they in effect providing loan guarantees to the Supply System? Bert Metzger, attorney with the Supply System's special counsel firm, later insisted the utilities had bought electricity. The utilities "were not in the

---

[61] "Washington Public Power Supply System Nuclear Projects Nos. 4 and 5 Participants' Agreement," Section 17, pp. 47–49, Exhibit 130/2 in Notebook of attorney Martha Walters, lent to author.

[62] The literature on the political economy of nineteenth-century developmental borrowing is extensive. Major sources include Carter Goodrich, *Government Promotion of American Canals and Railroads* (New York: Columbia University Press, 1960); Louis Hartz, *Economic Policy and Democratic Thought: Pennsylvania, 1776–1860* (Cambridge: Harvard University Press, 1948), chapter 4, "The Public Works"; Eric Foner, *Reconstruction: America's Unfinished Revolution: 1863–1877* (New York: Harper and Row, 1983), especially 379–392; Mark W. Summers, *Railroads, Reconstruction, and the Gospel of Prosperity* (Princeton, NJ: Princeton University Press, 1984); David Thelen, *Paths of Resistance* (first published 1986; Columbia, MO and London: University of Missouri Press, 1991), 62–70. A. M. Hillhouse, *Municipal Bonds: A Century of Experience* (New York: Prentice-Hall, 1936) remains the most comprehensive source on state and local issues. For estimate on defaults in the 1870s, see Hillhouse, 39.

banking business; they were in the power business."[63] However, when WPPSS terminated the plants in 1982, participants eager to avoid paying for power they never would receive challenged the agreements as loans of credit to WPPSS.

Houghton Cluck Coughlin and Riley, the Seattle law firm that served as special counsel to the Supply System, and Wood Dawson Love and Sabatine, WPPSS's New York bond counsel, led the effort to determine the validity of the Participants' Agreements between the preference customer utilities and the Supply System. Both firms had extensive ties to public power. Jack Cluck, senior partner in the Seattle firm, had handled condemnation proceedings which Public Utility Districts in Washington had used to acquire private utility facilities. He and his firm had served Northwest public power interests since the early 1930s. Public power was more than a client; for Cluck it was a noble social cause.[64] In New York, Wood Dawson specialized in representing municipalities and public agencies as bond counsel, serving public power utilities and other issuers around the country. The counsels' specialized expertise no doubt bolstered utility executives' judgment that they had cleared the legal hurdles facing utility participation in WPPSS Plants 4 and 5.

The question of whether municipalities and PUDs would be illegally lending their credit to the Supply System had surfaced initially in late 1973 during discussion of preliminary financing for WNP-4. At that time, both Houghton Cluck and attorneys for the potential participants worried about legal roadblocks. Jack Cluck wrote in November 1973 that Bonneville preference customers which were not themselves WPPSS members could not lend money to the Supply System. Nor could they legally promise to repay money the System had borrowed elsewhere. The agreements might constitute an improper loan of credit. Thus, Cluck commented, "Note financing by Supply System based on loan guarantees of municipal preference customers which are not Supply System members is not available because of legal restrictions."[65]

According to Metzger, he and Jack Cluck examined "everything we could find" on the question of whether Northwest utilities had the authority to enter into take-or-pay agreements with WPPSS and concluded they would be valid. However, the 1988 Securities and Exchange Commission

---

[63] SEC, *Staff Report*, 304.
[64] Billington, *People, Politics & Public Power*, 452–455, offers a sketch of his friend Jack Cluck.
[65] SEC, *Staff Report*, 304.

Staff Report investigated this claim and expressed skepticism, noting, "No memorandum exists setting forth the research conducted, the conclusions reached, and the basis for those conclusions."[66] There were no court cases in Oregon, Washington, or Idaho that set precedents on whether municipal authorities could sign such contracts.

Wood Dawson, the Supply System's bond counsel, also failed to find regional precedents for utility take-or-pay contracts. However, attorney Brendan O'Brien also told the SEC Staff Investigation that Wood Dawson had been confident of the pacts' validity. His review of the Participants' Agreements concluded that the powers of municipal bodies and PUDs were broad enough, and that the Supply System and the participants were on firm ground in these contracts.[67]

In reaching these optimistic conclusions, the lawyers seem to have put aside some of their worries about utilities' loans of credit that had surfaced during discussions of preliminary financing in 1973. In the spring of 1975, they decided not to pursue a test case on whether municipalities in Idaho had the authority to enter into option contracts. Though "very confident as to the likely success of such litigation," project proponents feared it would cause costly delays. They also seemed afraid that anti-nuclear or environmentalist interveners might drag out any test litigation.[68] When the Option Agreements were executed in July 1975, cryptic attorneys' notes indicated that they felt the agreements were valid because the utilities were entering a bilateral contract with mutual obligations. The utilities agreed to pay off their share of the bond obligations; the Supply System agreed in exchange to reserve a share of the projects' capability and to provide "transmission, scheduling, load factoring, reserves, exchanges and other services" to the utilities in conjunction with the projects. This, they hoped, would differentiate the Option Agreement from a unilateral loan guarantee by the signing utilities.[69]

Anxieties about the validity of take-or-pay contracts re-emerged in dealing with the Direct Service Industries. As noted above, bringing WNP-4 and 5 on line in the early 1980s meant the plants would be generating electricity before the participating utilities needed their project shares. Short Term Sales Agreements covering 1982 through 1988 (with possible

---

[66] Ibid., 297, 295.   [67] Ibid., 299.

[68] Norman A. Stoll, Public Power Council attorney, to R. Ken Dyer, Public Power Council General Manager, April 4, 1975, p. 4, Exhibit 183/128 in files of attorney Martha Walters, lent to author. SEC, *Staff Report*, 326–328.

[69] Ibid., 308–310; "Option and Services Agreement", p. 3. Exhibit 129/1 in files of attorney Martha. Walters, lent to author.

extensions to 1992) would dispose of this surplus to the Direct Service Industries. However, the companies balked at the first version of the Short Term Sales Agreements. The companies complained that the draft contracts might require them to pay even if the reactors failed to deliver electricity between their start-up and 1988. Ultimately, in late 1976, the Short Term Sales Agreements were revised to make them more clearly contracts for sales of electric power. The DSI snag and the Sales Agreement revisions it required had been "discouraging and upsetting to the Supply System and the Participants; and, the Companies have expressed their sincere apologies for the inconvenience it causes us all."[70]

The flap over these agreements indicates how Northwest energy leaders escalated their commitments to a strategy of generating supply growth. The Supply System and Bonneville initially justified WNP-4 and 5 in order to meet future energy needs of public utilities in the region. Yet in 1976 WPPSS executive H. R. (Hank) Kosmata was explaining to the project participants that the Short Term Sales Agreements were now crucial because "even if the load forecasts of the Participants were to significantly drop, the STSA and Assignment Agreements [by which public utility participants would transfer allotments to the Direct Service Industries until the publics needed the power] provide a mechanism for protection from the possibility of large cost exposure for surplus capability." Underwriters and bond buyers would demand such assurance.[71] In effect, Kosmata was saying that the projects should go forth whether or not there would be shortages without them; the goal now was to build the plants, even if they might not be needed. Only this (generally unconscious, it appears) shift in purposes seems capable of explaining why nuclear power supporters fretted so much about delays in completing plants. Unfortunately, this displacement of objectives, from providing service to building capital-intensive generating facilities, became a motif in the history of the Supply System.

Beset by legal and policy complications, yet convinced the projects were crucial, the Washington Public Power Supply System decided in the spring of 1976 to submit Participants' Agreement contracts to 93 public utilities in the region. Bonneville kept exhorting utilities to sign on for WNP-4 and 5. Some of the smaller utilities still believed that, as requirements customers, they would receive all the power they needed from Bonneville without participating in the new projects. BPA countered their reluctance

---

[70] SEC, *Staff Report*, 310–313; H. R. Kosmata to Participants in WNP 4/5, 22 December 1976, p. 1. (Exhibit 173/45 in files of attorney Martha Walters, lent to author.)
[71] Ibid., p. 2.

to sign with warnings and promises. One utility lawyer reported that Bonneville officials told him that the agreements had to be signed without any revisions; "any changes...would be unacceptable and would jeopardize the ability for [his clients] to participate in the project."[72] BPA informed preference customers that it would not use participation in WNP-4 and 5 as a rationale to decrease a utility's hydropower allocation in the future.

While utilities around the Northwest deliberated on the new projects, Donald Hodel, on June 24, 1976, issued the long-anticipated Notice of Insufficiency. From July 1983 on, Bonneville would not be able to meet the power demands of its preference customers and would have to apportion its limited supply among them. Since most of the preference utilities had no generating capacity of their own, the Notice carried a powerful incentive to buy into a share of the resource that the Supply System claimed would come on line just when Bonneville's power was to be curtailed. Heightening the pressure on the preference customers was the fact that Bonneville also was warning it could not offer its DSI customers the same level of assured, uninterruptible power when their contracts came up for renewal in the 1980s.[73] If the DSIs could not buy their electricity from Bonneville, they would turn to the utilities serving the areas in which they were located. Eighty-five percent of them were in preference customer territory. Not surprisingly, Bonneville preference customer utilities were inclined to see the chance to participate in the new WPPSS plants as an offer they could not prudently refuse.

Two Utilities Decide

By looking at the experience of the most notable utility that declined participation, Seattle City Light, and at the case of the Springfield Utility Board, which took almost 2 percent of the projects' capability, we can see the Pacific Northwest poised between its long-standing belief in energy-led growth and a newer, more skeptical approach.

Seattle City Light, the largest municipal system in the Northwest, held (and still holds) a unique position among the region's public utilities.

---

[72] Gleckman, "WPPSS: From Dream to Default," 161.

[73] In a statement to a People's Utility District being formed in Oregon, BPA stated, "Under our current load-resource assumptions, industrial customers would receive a significantly lower class of power. Nearly all of this power could have been [sic] restricted for various reasons." U.S. Bonneville Power Administration, "Statement of Bonneville Power Administration concerning Availability of Electric Power and Services for Proposed Emerald Peoples' Utility District," reprinted in *Bonneville Financial Fallout*, 249.

Although it had become a member of WPPSS in 1971, it produced 70 percent of its own power and depended less on Bonneville's energy than most other Supply System members. Nevertheless, when BPA and WPPSS offered Option Agreements for Projects 4 and 5 in early 1975, City Light managers wanted to join in. Predicting a continuing load growth of 3 to 4 percent annually, the utility asked the City Council to authorize purchase of an option on 10 percent of the plants' capability. Local business interests and media endorsed the proposal; the Council unanimously approved it in May 1975.

When the Washington Environmental Council claimed that Seattle's involvement in the new nuclear projects required an Environmental Impact Statement and sued to block the City Council's action, officials from the utility and the mayor's office negotiated an agreement with the environmentalists that called for an extensive analysis by an independent consultant. This became the Energy 1990 study, a pioneering attempt to cast community electrical energy needs in a broad framework.[74]

A growing drive for citizen participation meant that Energy 1990 underwent intensive public scrutiny. A Citizens' Selection Committee took part in choosing an independent consultant group to conduct the study. A Citizens' Overview Committee oversaw the consultants' work. Urban historian Carl Abbott has effectively contrasted Seattle's growth-oriented planning with Portland's cautious commitment to process and participation, but in the case of Energy 1990 public involvement was high. Nearly 7,000 residents completed a detailed questionnaire expressing their views on City Light's policies. For 38.2 percent of the respondents, "impact on environment" was the most important consideration in energy planning, while only 20.9 percent considered "impact on local economy" to be the top priority.[75] Meanwhile, both the Mayor's Office of Policy Planning and Seattle City Light itself contained environmentally minded professionals who doubted facile assertions that participation in WNP-4 and 5 was necessary.[76] Crucially, Mathematical Sciences Northwest, the consulting

[74] Two excellent studies of the Seattle experience with WNP-4 and 5 are: Joseph Gregory Hill, *The Public Interest and the Evaluation of Public Policy*, Ph.D. Dissertation, University of Washington, 1981, especially ch. 1, "The Politics of Energy 1990"; Wayne Sugai, "The WNP 4 & 5 Decision: Seattle and Tacoma – A Tale of Two Cities," *Northwest Environmental Journal* 1 (1984): 45–95.

[75] Hill, *Public Interest and the Evaluation*, 41–42.

[76] Carl Abbott, "Regional City and Network City: Portland and Seattle in the Twentieth Century," *Western Historical Quarterly* 23, 3 (August 1992): 293–322; Seattle Department of Lighting, *Energy 1990: Final Report*, May 1976, v.3, pp. 3b-83, 3b-86.

firm chosen for the forecasting component of Energy 1990, combined environmentalist values with its technical expertise. Math Sciences Northwest challenged some of the key assumptions behind the nuclear option. It produced the region's most sophisticated demand forecast. Paradoxically, although they used advanced statistical techniques, the consultants stressed the ambiguity of demand prediction. Whereas conventional forecasting methods usually assumed that utilities would respond with generating facilities to serve load levels they could not control and calculated demand as a single value, newer forecasting methods posited that utility policies themselves could affect demand, that generating more kilowatts was not the only way to fulfill customer needs, and that forecasts should predict ranges of likely outcomes, not single values, for demand at future dates. In the initial report of Energy 1990, issued in February 1976, the authors presented seven different energy futures. The report frankly stated, "[T]he scenarios should be regarded as literary tools." The new forecasting methods, despite their recognition of uncertainty, clearly challenged assumptions of unceasing demand growth. Whereas Seattle City Light had predicted that demand would grow by an average of 3.72 percent annually through 1990, the most likely rate according to Mathematical Sciences Northwest was only 1.52 percent.[77]

One major difference between the utility and the consultants stemmed from City Light's assumption that excess power from the Supply System plants could be sold to other utilities at its full cost. The consultants doubted that the market in the 1980s would be so strong. If Seattle lost money disposing of its surplus, its own rates would likely rise and customers would thus demand less electricity than with full-cost surplus sales. This seemingly technical forecasting dispute reflected the theoretical and ideological presuppositions of proponents and foes of rapid electrical supply growth.

The initial Energy 1990 Report appeared at the end of February 1976, as regional power leaders were intensifying the drive for participation in WNP-4 and 5. Its seven energy policy scenarios ranged from aggressive promotion of electricity usage to measures designed to reach a steady state with no growth in demand. The report made clear that investing in more thermal generating facilities was unappealing: "There appear to be no truly attractive central-station generation options available to Seattle."[78]

---

[77] Seattle Office of Environmental Affairs, *Energy 1990 Study: Initial Report*, v.1, February 1976, p. 7-3, pp. 3-8, 3-13.
[78] Ibid., v.1, pp. 4-15.

In the following months, Seattle residents witnessed, and to a surprisingly large extent took part in, an intense debate about the city's energy future. Public hearings began in March and drew testimony and submissions from a wide variety of groups. The City Council held a series of briefing sessions on energy policy. Seattle newspapers covered the issue regularly and televisions stations devoted specials to it.

Late in April 1976, the Citizens' Overview Committee transmitted its majority and minority reports. Eighteen members signed a statement calling for a vigorous conservation program and recommending "that no new additional generation be initiated at this time." The remaining nine filed a minority report accusing the majority of playing a "dangerous energy game" and endorsing City Light's purchase of 10 percent shares of WPPSS 4 and 5 capabilities.[79]

When the final Energy 1990 Report appeared in May, City Light Superintendent Gordon Vickery transmitted it to Mayor Wes Uhlman with the utility's recommended course of action, purchase of a 5 percent share in the WPPSS projects and additional investments in hydropower and coal generation. Uhlman accepted the nuclear aspect of the proposal but added a proviso that participation should be "contingent upon the identification of specific customers which can reasonably be expected to purchase any surplus power which might accrue to us."[80]

City Light's 5 percent plan won the backing of an alliance of downtown business interests, labor unions, construction contractors, Seattle's daily newspapers, and the region's private utilities. Perhaps equally predictably, the opponents of nuclear involvement "formed a classic progressive coalition – environmentalists, academics, community councils, the League of Women Voters, People Power, Metrocenter [neighborhood and consumer groups] and the Municipal League – the sort of middle-class groups that were coming to exercise more and more power" in Seattle, according to political scientist J. Gregory Hill.[81]

As Hill points out, the pro-nuclear forces were not as invincible as their economic importance might suggest. Opponents of participation

---

[79] Seattle City Light, *Energy 1990: Final Report*, May 1976, Part B, pp. 3b–17, 3b–23.

[80] Mayor Wes Uhlman to City Council, May 20, 1976, p. 2, copy in Energy 1990 Notebook, Seattle Municipal Research Library.

[81] Hill, *Public Interest and the Evaluation*, 66–67. For a broad overview of value change and political mobilization in late twentieth-century America and Western Europe, see Ronald Inglehart, *The Silent Revolution: Changing Values and Political Styles among Western Publics* (Princeton, NJ: Princeton University Press, 1977).

portrayed themselves as fiscally responsible moderates backed by technical expertise and objective information. Moreover, they cast themselves
as the legatees of the democratic impulses of the public power movement
and of City Light's own progressive tradition. (Ironically, as we have seen,
WPPSS itself and the pro-nuclear public utility interests also laid claim to
the region's public power heritage.) The foes of nuclear investment combined these themes with a kind of civic patriotism because advocates of
nuclear participation implied that taking shares in the Supply System's
projects was an obligation Seattle had to the region and beyond. The
anti-nuclear side thus could claim to be the ones who kept Seattle residents' interests in the foreground.

The City Council delayed its vote on WPPSS 4 and 5 participation
as long as possible, but the Participants' Agreements had to be returned
by mid-July. On July 12, the Council voted 6–3 against acquiring the
5 percent shares. They also rejected, 7–2, a proposal for a 1 percent share
designed to keep the city's foot in the door for future nuclear projects.
An alternative vision of energy policy had prevailed in the Northwest's
largest city.

Developments in Springfield, Oregon, were less dramatic but far more
typical of the responses of local utilities to the nuclear offer that WPPSS
and Bonneville were making. The Springfield Utility Board (SUB) had
begun distributing electricity to the rapidly growing blue-collar city in
1949. Although in its earlier years it had competed actively for customers against investor-owned Pacific Power & Light, SUB's culture developed in the 1950s, with the Eisenhower era's emphasis on "partnership,"
rather than the earlier years of bitter public-private utility rivalries. Unlike
Seattle, Springfield had no strong tradition of citizen participation or a
sense of public power as a progressive crusade. Minutes and newspaper
coverage indicate that few if any citizens attended most Board meetings.
Board decisions were usually unanimous, following staff recommendations, and discussion seems seldom to have been more than perfunctory.
The city's semi-weekly newspaper, the *Springfield News*, routinely covered SUB meetings, but reports were more likely to note a decision to
purchase a new truck or provide service to a new housing tract than to
discuss impending policy choices.

Thus, Springfield signed its Option Agreement in July 1975 for the
opportunity to buy a bit less than 2 percent of the capability of Projects
4 and 5 with virtually no Board discussion and no mention of this move
in the local press. BPA's announcement that it would issue notices of

insufficiency drew scant attention in the spring of 1976. When it came time in July 1976 for the Utility Board to sign its Participant's Agreement for WNP-4 and 5, there was slightly more deliberation. At a work session on July 12, SUB General Manager Jack Criswell presented a forecast showing a fourteen megawatt deficiency in 1983–84 even with vigorous conservation efforts. Two days later, at the Board's monthly meeting, three citizens spoke against the Participant's Agreement. One questioned Criswell's demand projections and called for a no-growth energy policy. Another opposed nuclear energy as experimental and noted the unsolved problems of waste disposal. A third announced that he was designing an invention that would generate electricity at a tenth of its current cost; details would soon be announced.[82]

The SUB Board, unpersuaded by these voices, was resigned to executing the Participant's Agreement. They saw no alternative if Springfield was to meet its customers' requirements in the next decade. Jack Criswell reassured one board member by reminding him of plans for short-term sales of any surplus power, but another noted that he was voting for signing the Participants' Agreement "reluctantly."[83]

Springfield's situation contrasted in almost all respects with Seattle's. City Light owned hydroelectric facilities that generated most of its power; SUB relied on Bonneville for all of its electricity. Seattle's population declined nearly 9 percent from 1970 to 1975, whereas Springfield's had grown by almost a quarter in the same years. Springfield lacked the tradition of citizen involvement and the sense of mission that public power in Seattle had inherited. Moreover, the communities themselves were strikingly dissimilar. Seattle, despite the class stratification to be found in any American city its size, was in some very real senses a middle-class city; Springfield, less than a tenth Seattle's size, was predominantly a blue-collar mill town. Seattle's per capita income was 38 percent higher than the Oregon community's. Almost two-thirds of its adult residents were high school graduates; only about half of Springfield's were. It was unlikely that the coalition that kept Seattle from signing the Participants' Agreements could have been duplicated in the Oregon community.[84]

[82] Minutes, Springfield Utility Board, April 14, 1976, p. 2, and July 14, 1976, p. 2; *Springfield News*, July 14, 1976, p. B1 and July 16, 1976, p. 1. *Eugene Register-Guard*, July 15, 1976, p. 3C.

[83] Minutes, Springfield Utility Board, July 14, 1976, p. 2.

[84] U.S. Bureau of the Census, *County and City Data Book: 1977* (Washington, D.C.: GPO, 1978), 734, 770.

Conclusion

When the Supply System's Board of Directors met on July 23, 1976, it received the happy news that eighty-eight utilities had returned signed contracts, and that they had, collectively, requested shares totaling 133 percent of the full capacity of WNP-4 and 5.[85] (As the Agreements provided, participant shares were prorated to equal 100 percent of project capability. Thus, a utility asking for 4 percent of project capability would have received a 3 percent share.). After two years of uncertainty, WPPSS had agreements to meet energy needs in the next decade and beyond. This would show public power's ability to undertake major regional projects and would solidify WPPSS's position as a leader in generating nuclear power, the energy source of the future.

Yet the situation in 1976 was full of portents of trouble. Three basic questions lay unresolved. First, could the Supply System realize its plans? Its nuclear empire was to become one of the largest under construction in the United States in the late 1970s, when WPPSS became the nation's largest issuer of municipal bonds.[86] Did the Supply System have the organizational and financial capabilities to fulfill its commitments? Ominously, in August 1976, little more than a month after the Participants' Agreements were returned, the Supply System announced a $540 million increase in anticipated costs of Projects 4 and 5. Escalation on the net-billed plants made the total cost estimates for all five projects almost a billion dollars higher than the previous predictions.[87]

Second, what were the consequences of failure? Would the lights go out and the machinery grind to a halt? Who would be legally responsible? The end of net billing, the collapse of Phase II, and the absence of a replacement regional energy program that would have allowed Bonneville to back up the WPPSS ventures on Plants 4 and 5 meant that the Supply System was starting across this tightrope with the flimsiest of legal and financial safety nets. Who would be injured if there were to be a fall? From the time of the Supply System's first bond issue for WNP-4 and 5, one indication of the precarious legal status of financing appeared in some curious wording in

[85] Minutes, WPPSS Board of Directors, July 23, 1976, p. 5.

[86] In 1978, for instance, of the eight largest competitively placed new municipal bond issues in the United States, six came from WPPSS. See Public Securities Association, *Statistical Yearbook Municipal Securities Data Base: The New Issue Market in 1978* (New York: Public Securities Association, 1979), Table L-1.

[87] Gleckman, "WPPSS: From Dream to Default," 165.

the bond sales' official statements: "We have examined into the validity of seventy-two of the Participants' Agreements," they noted, and went on to say they considered these contracts valid. However, all eighty-eight participants had signed agreements. Both law firms, Houghton Cluck and Wood Dawson, had declined to endorse sixteen of them, in most cases because they feared the utility lacked legal authority to enter into a contract for project capability. All sixteen participants were small; their collective share came to less than 4 percent of the projects' expected output. Unfortunately for the Supply System and for the region's nuclear enthusiasts, the lawyers' doubts about these sixteen were only a hint of the legal complications to come after WNP-4 and 5 were terminated in 1982.

Third, were WPPSS, Bonneville, the Public Power Council and others who had pressed for these capital-intensive, large-scale thermal generating facilities, right about demand? Were these plants necessary? Already, the Eugene nuclear referendum, the Seattle Energy 1990 study, and the environmentalist pressures leading to the Alumax decision and Bonneville's Role Environmental Impact Statement were pointing to different paths for Northwest energy policy. Nationally, the pace of electrical generation growth was slowing sharply. Between 1963 and 1973, net generation more than doubled. In the following decade, it rose by only 24 percent.[88] At a moment when patterns of energy supply and demand were undergoing sharp changes in direction, the Washington Public Power Supply System had embarked on a perilous course.

---

[88] U.S. Department of Energy, Energy Information Administration, *Annual Energy Review 2002*, Table 8.1 Electricity Overview, 1949–2002, http://www.eia.doe.gov/emeu/aer/txt/ptb0801.html, accessed July 20, 2004.

4

The Construction Morass

It is easy but not altogether useful to mock the aspirations of nuclear power's promoters in their heyday. Electricity too cheap to meter, power plants without pollution, the alchemy of the breeder reactor – these all ring hollow today, three decades after the last reactor order was placed. Unexpected forces have buffeted the once-staid electrical industry. New paradigms and panaceas appear, and the brave new atomic future looks to many like yesterday's distraction and today's burden.* When nuclear energy makes headlines today in the United States, the topics often are decommissioning, accidents, malfunctions, or the "stranded costs" of generating facilities too expensive in an era of intensified competition. Hindsight, however, may blind us to the reasons why nuclear power attracted so many in the 1960s and 1970s or lead us to accept oversimplified accounts of the demise of that era's nuclear dream. Nor will it provide answers to the very real energy dilemmas we face today.

Those who are not specialists tend to explain the decline of nuclear power in the United States by two vivid events: the 1979 accident at Three Mile Island, Pennsylvania, and the 1986 Chernobyl disaster. There is no denying their importance. Public opinion polls showed a qualitative shift away from support of nuclear power after Three Mile Island.[1]

---

* Of course, reports of the demise of nuclear power are greatly exaggerated. Nuclear reactors are second only to coal plants as sources of electricity in the United States. As noted in the preface, a new push to resume nuclear construction is underway.

[1] Stanley M. Nealey, Barbara D. Melber, William L. Rankin, *Public Opinion and Nuclear Energy* (Lexington, MA: D.C. Heath, Lexington Books, 1983), 31. See also William R. Freudenburg and Eugene A. Rosa, eds., *Public Reactions to Nuclear Power: Are There Critical Masses?*, American Association for the Advancement of Science Selected Symposium 93 (Boulder, CO: Westview Press, 1984).

In 1986, the human and environmental damage spread through a broad region, alarmed the citizens of Europe, east and west, and shocked the globe. Yet other nemeses of nuclear power in the United States had made their presence known before these episodes. Electric demand forecasts had proven to be gross overestimates. Large-scale nuclear reactors had already called forth large-scale and spirited popular resistance, ranging from legal challenges to mass civil disobedience. Technical schemes for dealing with radioactive waste disposal were untested and politically unpalatable. The nuclear power industry had not succeeded in shaking its implicit, and sometimes explicit, identification with nuclear weaponry. Finally, American utilities had seldom managed to build nuclear plants on time and within cost projections. Even before Three Mile Island, nuclear power looked to many like part of the problem, not the solution, of high energy costs.

This chapter explains how the WPPSS nuclear plants fell years behind schedule and why their projected costs climbed about 400 percent in less than a decade. Of course, the American utility landscape is studded with nuclear plants that began operations years late and billions of dollars over cost or never operated at all. The Supply System's story cannot be understood apart from the broader fate of nuclear power in the years of both its brightest hopes and its gravest problems. Two other points also deserve mention. First, the problems in nuclear construction were not exclusively American, but they were acute here. President Nixon's proposal for "energy independence" that entailed generating half of the nation's electricity in nuclear plants by 2000 never came close to realization. In contrast, France – the nation with the clearest governmental commitment to nuclear power – gets over three-quarters of its electricity this way.[2] In the mid-1970s, France, notably lacking in fossil fuels, brushed aside (sometimes forcibly) substantial popular opposition and elite discord to institute a program of standardized plants that have provided that nation with electricity at a socially acceptable price.[3] Second,

---

[2] International Atomic Energy Agency, "Country Nuclear Power Profiles," France, 2003, http://www-pub.iaea.org/MTCD/publications/PDF/cnpp2003/CNPP_Webpage/countryprofiles/France/France2003.htm, Table 6, accessed July 23, 2004.

[3] On the French case, see the insightful comparisons in James M. Jasper, *Nuclear Politics: Energy and the State in the United States, Sweden and France* (Princeton, NJ: Princeton University Press, 1990) and Jasper, "Rational Reconstructions of Energy Choices in France," in James F. Short, Jr., and Lee Clark, eds., *Organizations, Uncertainties, and Risk* (Boulder, CO: Westview Press, 1992), 223–233. Another illuminating approach is Gabrielle Hecht, *The Radiance of France: Nuclear Power and National Identity after World War II* (Cambridge, MA: MIT Press, 1998).

we should avoid the fatalism that views the American industry's problems as inevitable and universal. LILCO's Shoreham in New York, Public Service Company of New Hampshire's Seabrook, and WPPSS are among the egregious cases of costly failures, but Duke Power Company in the Carolinas was one of several utilities with a generally positive record of meeting construction goals.[4]

Although emphases differ, most analysts agree on several factors behind the nuclear project travails of the 1970s and 1980s. The atom's advocates emphasized cumbersome and frustrating licensing and regulatory procedures. Utilities had to get a construction permit from the Atomic Energy Commission or its successor, the Nuclear Regulatory Commission (NRC). Then, after investing years and at least several hundred million dollars to build the plant, they had to apply to the agency for an operating license. At each stage, intervenors could – the utilities maintained – quibble, obstruct, and delay those who wanted to supply users with needed energy. Compounding these problems, states could regulate the siting of nuclear plants. State siting procedures could also bring out opposition from siting council members and staff, environmentalists, and "NIMBY" ("Not In My Back Yard") groups averse to having a nuclear plant located near them.

Another common explanation of nuclear construction problems, one less flattering to the nuclear industry, is that American utilities and nuclear construction companies failed to develop a standard model of an efficient, reliable nuclear plant. Unlike France, the United States lacked a nuclear plant design to take off the shelf. Each project required an architect-engineering (A/E) firm to devise its reactor design and associated structures. Learning by doing, which might reduce costs over time, advanced more slowly when each plant was unique. Although there were cost savings for plants designed by more experienced A/E firms, the advantages were modest. Related to this, utilities frequently built nuclear plants "ahead of design." Construction would commence without complete blueprints, and planning would continue as the structures arose. This fast-track construction is common in large projects, but the fast track has its perils. These projects often find themselves reversing course, re-doing or undoing tasks already completed.

The electrical utility industry in general, and nuclear power in particular, put great stock in building ever-larger facilities. As we have seen, the history of utility generation facilities had, through the 1960s, taught the lesson that bigger meant cheaper, that economies of scale were available

---

[4] See, e.g., "The Best," *Forbes*, February 11, 1985, 93.

to those who could build larger plants. Owen Hurd, the Supply System's first Managing Director, expressed this view as early as 1958, in a speech to the American Public Power Association: "The economics of electric power generation and transmission inescapably lends itself to bigness . . . It is becoming more apparent that economic nuclear power of the future will be from plants of large capacity located away from centers of population."[5]

However, by the end of the 1970s, hardheaded examinations of power plant costs looked bleak for nuclear power. In 1981, Charles Komanoff, an energy economist in New York, completed a damning comparison of nuclear and coal capital costs, *Power Plant Cost Escalation*. Contrary to utility industry calculations, Komanoff found that nuclear projects completed in the 1970s cost over 50 percent more than coal plants per megawatt of capacity. Moreover, the tide was moving against nuclear. As the number of reactors increased, even nuclear proponents like physicist Alvin Weinberg contended that the risk of a serious accident at each individual reactor would have to decrease to keep the nuclear industry within an acceptable margin of safety. Thus, safety concerns would impose increasingly costly regulations on new projects. Komanoff projected that capital costs of plants completed in 1988 would be 73 percent above coal plants of the same vintage.[6]

As the Washington Public Power Supply System undertook to build, first, the three net-billed plants and, thereafter, the two projects backed by utilities' Participants' Agreements, it encountered virtually all the plagues commonly visited upon the nation's nuclear program. In addition, the consortium faced problems distinctively, if not uniquely, its own. Some of these difficulties, in turn, stemmed from the legal and social environment

[5] Owen W. Hurd, "Cooperating with Other Organizations to Strengthen Public Power," American Public Power Association Convention, New Orleans, May 6, 1958, p. 3. Copy in [Hurd], *A Collection of His Speeches*, loose-leaf volume at Washington Public Power Supply System Library, Richland, WA.

[6] Charles Komanoff, *Power Plant Cost Escalation: Nuclear and Coal Capital Costs, Regulation, and Economics* (New York: Van Nostrand Reinhold, 1981), passim; also Komanoff, "Nuclear Costs Spiral above Coal," *Public Power* 39, 5 (September-October 1981): 70ff; Richard Hellman and Caroline J.C. Hellman, *The Competitive Economics of Nuclear and Coal Power* (Lexington Books: Lexington, MA, 1983); Irvin C. Bupp and Jean-Claude Derian, *The Failed Promise of Nuclear Power: The Story of Light Water* (New York: Basic Books, 1981). Weinberg's case that safer nuclear plants were needed is in Alvin M. Weinberg, "An Acceptable Nuclear Future?" *Sciences* 17, 8 (December 1977), 18–23. See also Weinberg, Irving Spiewak, Jack N. Barkenbus, Robert S. Livingston and Doan L. Phung; Russ Manning, Editor, *The Second Nuclear Era: A New Start for Nuclear Power* (New York: Praeger, 1985), 3–25.

in which the Supply System operated. However, many of the woes WPPSS faced were of its own making. This was not an organization with the capacity to complete these projects efficiently and punctually. Organizational failure was a major motif in the story of the WPPSS nuclear projects.

Management Weaknesses

A glance at the Washington Public Power Supply System before it embarked on its reactor projects might—with hindsight, at least—suggest that this agency was not prepared to undertake a multi-billion dollar construction program. In 1971, the year before it broke ground for its first nuclear plant, WPPSS had only a few dozen employees.[7] Its generating assets consisted of its small hydroelectric plant at Packwood Lake in central Washington and the generating facilities attached to the Hanford N-Reactor. Structurally, each of the member utilities, most of them from small and medium-sized communities around Washington, had a representative on the WPPSS Board of Directors; each member had one vote regardless of size. The Board chose a five-member Executive Committee. Until a series of administrative reforms during the crisis of the early 1980s, nobody except utility representatives sat on either body, and the Board itself selected all the Executive Committee members. Most Board members were Public Utility District commissioners, not utility professionals. The two largest municipal utilities in the state, Seattle City Light and Tacoma City Light, did not even join the Supply System until 1971 and 1972, respectively.[8] The System's member utilities themselves had hardly been hotbeds of innovation in recent decades. Since most WPPSS members relied on Bonneville for all of their power, they rarely possessed experience with major construction projects.

Minutes of Board and Executive Committee meetings reveal a rather narrow perspective. Not only does the Supply System's structure appear inadequate to the huge undertakings it assumed, its processes also seem inappropriate for its daunting tasks. Meetings devoted time to authorizing managers' travel to conferences in the region, reports on study programs

---

[7] Sources provide slightly varying employment numbers. The figure of eighty-one comes from Coopers & Lybrand, Certified Public Accountants, "Review of Contract Administration and Project Accounting," Final Report, May 17, 1978, p. 1.

[8] Minutes, WPPSS Board of Directors, Vancouver, WA, March 23, 1971, p. 5; Minutes, WPPSS Executive Committee Special Meeting. Vancouver, WA, October 20, 1972, p. 1.

on nuclear energy for Northwest high school students, appropriations for vehicle repair, and the like. At its last meeting before Owen Hurd's retirement, the Board adjourned to examine Hurd's new motor home. Later that year, he returned to a Board meeting to report on his trip through Canada and his new home in Sun City.[9] Interspersed with these quotidian topics were discussions of commitments of tens or hundreds of millions of dollars and, only sporadically, consideration of broader topics. One consultant's study reported that, at a typical Executive Committee meeting, one lasting two and one-half hours, only three minutes had been devoted to policy issues.[10]

When Robert Ferguson took over as Managing Director in 1980, he was dismayed at Board and Executive Committee operations. They spent "enormous time...arguing over vehicles...the color of rugs...things they understood, and the big stuff went sailing through there...It was just incredible."[11] As for early management itself, the executive vice president of the Tri-Cities Nuclear Industrial Council, an improbable critic, later excoriated those in charge: "They wouldn't listen to people. They thought they knew it all when they didn't know a damned thing!"[12]

Top management reflected this pattern of provincialism. The Supply System's first Managing Director, Hurd, had headed the Benton County Public Utility District, itself headquartered in Richland. J. J. (Jack) Stein, Hurd's successor in 1971, had also been a PUD manager, in Grays Harbor County (which became the site of the now-abandoned nuclear plants three and five). Indeed, he was on the WPPSS Board and was initially named to a Candidate Selection Committee to choose a Managing Director to follow Hurd.[13] When Stein retired in 1977, the Supply System again chose from among its own, selecting Neil O. Strand, who had spent seven years in the ranks of management, most recently as Director of Projects. Strand was the first Managing Director with nuclear energy experience elsewhere, but this dated from the early 1960s. It was reported that Strand won the position in large measure because he was willing to accept a salary less

[9] Minutes, WPPSS Executive Committee, June 25, 1971, p. 3; Minutes, WPPSS Board of Directors, December 8, 1971, Seattle, p. 7.

[10] Cresap, McCormick and Paget, Inc., Management Consultants, *Washington Public Power Supply System Study of Management Organization and Related Issues*, August 1976, III-17.

[11] Robert Ferguson, interview by author, tape recording, Kennewick, WA, October 15, 1992.

[12] Joel Connelly and Don Tewkesbury, "Trouble-plagued WPPSS will run its first N-plant," *Seattle Post-Intelligencer*, August 17, 1984.

[13] Minutes, WPPSS Board of Directors, June 26, 1970, p. 5.

than half of what a management consultant recommended for a position of this magnitude.[14] In 1980, as the agency's projects were drifting out of control, the Board dismissed Strand and hired Robert Ferguson. The former Deputy Assistant Secretary of Energy for nuclear programs, Ferguson was the first Managing Director to assume the position with a record of accomplishment at the national level. However, by that time, even this highly regarded "no-nonsense" manager could not stanch the flow of dollars without curtailing the projects.

Supply System management delighted in reaping praise from consultants who sprinkled compliments to their clients through their reports. Thus, in 1978, a presentation from Cresap, McCormick and Paget lauded WPPSS for its "very good progress" in implementing management reforms the consultants had proposed in their 1976 study. "Clearly, WPPSS is moving in the direction of becoming a mature business and technical organization, and an effective representative of your collective interests."[15] In mid-1977, with delays and overruns already mounting, Don Patterson of the Supply System's financial advisor Blyth Eastman Dillon & Co. told the Board of Directors of "the Supply System's excellent reputation in the financial community and the positive effect this will have on future bond issues, particularly for Projects 4 and 5."[16] If external approbation was insufficient, the agency was not above resorting to self-compliments. (As Howard Gleckman, who covered the Supply System extensively for the trade publication *The Bond Buyer* noted, its public relations arm was far more sophisticated than the rest of the organization.[17]) Annual reports regularly touted the dedication, experience, and skill of the organization's leadership and its professional staff.

The realities were less flattering. Owen Hurd may have been a visionary leader and a charismatic advocate for nuclear energy development, but his reputation as an administrator was less than sterling.[18] Jack Stein

---

[14] Gary K. Miller, *Energy Northwest: A History of the Washington Public Power Supply System* (n.p.: Xlibris, 2001), 260 notes the executive search consultants' recommendation of a $150,000 salary. Strand appears to have been hired at less than $65,000. The Board's Executive Committee approved a raise to that amount three months after Strand's appointment: Minutes, WPPSS Executive Committee, Seattle, June 24, 1977, p. 10.

[15] Cresap, McCormick and Paget, Inc., "Washington Public Power Supply System Study of Management Organization and Related Issues: Review of Progress Toward Implementation July 1976-January 1978," in Administrative Central File, Washington Public Power Supply System Records Management Division, Richland WA.

[16] Minutes, WPPSS Board of Directors Special Meeting, Seattle, July 12, 1977, p. 2.

[17] Howard Gleckman, interview by author, telephone, December 3, 1997.

[18] Gleckman interview.

was regarded as competent and tough-minded, but was a "difficult man to work for, I'll tell you," recalled Richard Quigley, the Supply System's General Counsel in those years.[19] Neil Strand evoked affection from his co-workers, but observers suggest that Strand was beyond his depth as Managing Director of an organization trying to build five large nuclear projects in complex circumstances.[20] Former Representative Jim Weaver (D-OR), the Supply System's congressional nemesis, recounts that Strand came on one occasion to testify before Weaver's committee. As usual, he flatly rejected Weaver's accusations against WPPSS. But during a recess in the hearings, the Congressman encountered Strand in a hallway and the Managing Director broke into tears, begging Weaver to help him get out of the mess.[21] Officials of Seattle and Tacoma City Light, along with William Hulbert, General Manager of Snohomish County's large PUD, were said to be behind Strand's 1980 dismissal.[22]

Howard Gleckman speculates that had Robert Ferguson taken over the Supply System five years earlier, the story would have taken a different course, but this may put too much responsibility on the individuals who served as Managing Director. Frank McElwee, Assistant Director for Projects from 1977 to 1980, ran through the prior experience of Supply System senior managers in the nuclear industry and on large construction projects, concluding, "So the management talent was very, very thin."[23] The organization faced stiff competition for top-quality project managers and nuclear engineers from other utilities building nuclear plants. The Supply System's salaries were below industry standards in the 1970s. A 1980 compensation survey found that for five out of twelve top management positions, the Supply System paid far less than comparable utilities. At this point, for instance, Managing Director Strand received $80,000 per year. The next-worst salary in the survey was $93,000 and the median $173,065.[24] When a State Senate Committee's Inquiry into the *Causes of Cost Overruns and Scheduling Delays* presented its final report in 1981, its verdict on WPPSS management was harsh: "The Committee identified

[19] Richard Quigley, interview by author, Kennewick, WA, October 15, 1992.
[20] Gleckman interview.
[21] James Weaver, interview by author, Eugene, OR, October 20, 1997.
[22] Gleckman, "WPPSS: From Dream to Default," 175–176; Gleckman interview.
[23] Frank McElwee, interview by author, Kennewick, WA, October 14, 1992.
[24] Arthur Young & Company, "Report on Executive Compensation for Supply System Senior Executives," Prepared for the information of the Managing Director, at the direction of the Executive Committee of the Board of Directors, April 1980, pp. 1–2, 7.

a number of areas of management failure, each of which significantly contributed to the cost and schedule problems on the projects... The cumulative impact of these deficiencies leads the Committee to conclude that WPPSS mismanagement has been the most significant cause of cost overruns and schedule delays."[25]

How Projects Fail

It should be clear that major projects in virtually all fields, not just nuclear power, are fearsomely complicated ventures. Although technological optimism seems to be a common characteristic of those who prescribe techniques of project management, reviews of large-scale projects indicate that most of them have been and are completed behind schedule and over budget. As Peter W. G. Morris and George H. Hough conclude, "Curiously, despite the enormous attention project management and analysis have received over the years, the track record of projects is fundamentally poor, particularly for the larger and more difficult ones." Although a few scholars have found projects, mostly small ones, that came in under budget or on target, "In all other cases, representing some 3,500 projects drawn from all over the world in several different industries, overruns are the norm, being typically between 40 and 200 per cent."[26] Another project management expert notes, only half-jokingly, "Projects never ever go according to plan. There are always deviations, hesitations, and interruptions."[27] Approximately half a century of the systematic study of project management and the refinement of planning techniques has not repealed Murphy's Law. Nor should we underestimate the broad human propensity for unjustified optimism. Experiments by Daniel Kahneman, the 2002 Nobel Prize recipient in Economic Sciences, demonstrate this. Flaws in human decision-making – overestimating one's own capabilities, paying too little attention to competitors' plans, and

---

[25] Washington State Senate Energy and Utilities Committee WPPSS Inquiry, *Causes of Cost Overruns and Schedule Delays on the Five WPPSS Nuclear Power Plants*, 2 vols. (Olympia, WA, 1981), I:3–4. Hereafter cited as Senate Inquiry, *Causes of Cost Overruns*.

[26] Peter W. G. Morris and George H. Hough, *The Anatomy of Major Projects: A Study of the Reality of Project Management* (Chichester, U.K.: John Wiley & Sons, 1987), 7. A few years later, Morris observed that a data base he compiled in the early 1980s showed only a dozen out of 1,449 projects had come in on or under budget. Peter W.G. Morris, *The Management of Projects* (London: Thomas Telford, 1994), viii.

[27] Geoff Reiss, *Project Management Demystified: Today's Tools and Techniques* (2d ed., London: E & F. N. Spon, 1995), 20–21.

"anchoring" forecasts to the desired outcomes – can be mitigated but rarely eliminated.[28]

The Danish sociologist Bent Flyvbjerg and his colleagues have even suggested that deceit, not mere fallibility, is at the root of the persistent underestimation of project costs. Their study of 258 transportation infrastructure projects around the world found that almost nine in ten of them were underestimated. There was no tendency toward improvement over time, indicating no learning from prior mistakes. The authors conclude that "cost estimates used in public debates, media coverage and decision making...are highly, systemically and significantly deceptive."[29] In one important sense, however, this study found a better situation than the Supply System's nuclear plants. On average, actual costs were only 28 percent higher than initially estimated. The WPPSS nuclear plants' cost overruns were an order of magnitude larger.

The ideas of economist Oliver Williamson may aid our understanding of why high technology construction projects commonly encounter difficulties in staying within target budgets. Williamson looks at the transaction costs involved in making and enforcing agreements between independent organizations.[30] One condition likely to boost the cost of contracting is the presence of information asymmetries. When one party knows more than the other about the subjects covered in the exchange, the less-informed side may have to spend heavily to avoid being victimized by the side that holds the upper hand. When a utility embarks upon a major generating facility employing a relatively new and complex technology (as was the case with nuclear energy in the 1970s), the utility almost certainly will find itself at a technological disadvantage in dealing with suppliers and builders. The Supply System's structure and procedures widened the gap.

A second condition for difficulties and high transactions costs in contractual relations that Williamson notes is the need for investments in specific assets, capital goods that are not likely to be marketable elsewhere.

---

[28] Dan Lovallo and Daniel Kahneman, "Delusions of Success: How Optimism Undermines Executives' Decisions," *Harvard Business Review* 81, 7 (July 2003): 56–63.

[29] Bent Flyvbjerg, Mette Skamris Holm, and Søren Buhl, "Underestimating Costs in Public Works Projects: Error or Lie?" *Journal of the American Planning Association* 68, 3 (Summer 2002): 279–295, quotation at 290. For a fuller report, see Flyvbjerg, Nils Bruzelius and Werner Rothengatter, *Megaprojects and Risk: An Anatomy of Ambition* (New York: Cambridge University Press, 2003).

[30] For overviews of Williamson's approach, see Oliver E. Williamson, *Markets and Hierarchies: Analysis and Antitrust Implications* (New York: Free Press, 1975) and Williamson, "Transaction-Cost Economics: The Governance of Contractual Relations," *Journal of Law and Economics* 22, 2 (October 1979): 233–261.

As he points out, when the buyer calls upon the seller to make a product or employ a production method that only that specific purchaser can or will use, it creates a condition of "bilateral monopoly." Each party in the transaction has a strategic interest in winning favorable terms at the expense of market efficiency. Indeed, Williamson cites the construction of a specialized plant at a specific location as an example of these conditions. These conditions demand what he calls "relational contracting," agreements that extend over time and allow flexible adjustments of the terms on which the parties do business. These relational contracting agreements in turn entail heightened costs of monitoring the relationship, costs that rarely can be fully anticipated. (In fact, if transactions costs are too high under relational contracting, there will be a strong incentive for a merger, substituting administrative coordination for market transactions.)

These conditions apply to the Supply System. Legislation bound it, as we shall see, to "arm's-length" contracts rather than relational arrangements. Dealing with the contractors was at first primarily in the hands of the architect-engineers until 1978 when WPPSS assumed an ambiguous and ultimately costly shared responsibility with the A/Es for construction management. On large-scale construction projects, the costs of adaptation over time typically take the form of change orders, directives that modify or expand upon previous agreements. Since the parties to contracts do not have the capacity to predict all contingencies in advance, change orders are virtually inevitable. If not controlled and managed appropriately, they are likely to be extremely expensive. The Supply System's contracting arrangements and its procedures for handling change orders proved to be among the most vexatious problems facing the projects' managers.

The organizational theorist Charles Perrow has also provided important insights into the challenges of large-scale projects. In an influential 1967 article he classified the technological environments of organizations according to two variables, the number of exceptions that present themselves, and the availability of analytical techniques to handle them. Technologies entailing many exceptions which are not readily handled with formal pre-existing techniques are the most complex and uncertain. For these, the most appropriate "task structure" is "flexible, polycentralized." Discretion and decision-making power will be in the hands of field supervisors and line workers; groups will rely heavily on experience and feedback rather than prescribed rules; different work groups will be highly interdependent. If, in fact, building a nuclear power plant requires coping with frequent exceptions that are difficult to analyze, the Supply System was poorly suited to meeting these circumstances. The stringent legal requirements for arm's-length contracting, for example, decreased the likelihood of cooperative relations among interdependent work groups. Safety concerns in nuclear projects meant detailed regulations and rule-bound methods of dealing with exceptions when experience-based responses might have been more helpful. Troublesome relations between contractors and construction trades unions could hinder effective exercise of power and decision-making in the field. According to the State Senate Inquiry, central office control hindered the responsiveness of managers of the individual projects.[31]

Perrow's later work extended his analysis to system failure. "Given the . . . characteristics [of certain systems], multiple and unexpected interactions of failures are inevitable."[32] These "system accidents" are "normal," not anomalous. Perrow's classic study, *Normal Accidents*, examines systems ranging from dams to spaceships to spell out the perils of large-scale undertakings. Although the "accidents" that befell the Supply System's projects were not direct physical threats to the public, Perrow's account may help us understand why major failures occurred during plant construction. Systems subject to normal accidents rank high on two characteristics, interactive complexity and tight coupling. Building nuclear plants certainly involves complex interaction, especially under fast-track conditions. Concurrent design and construction entails nonlinear interaction of parts of the system. Construction incidents, for example, may require readjusting the design or engineering processes; engineering choices might have unexpected consequences for quality control or inspection procedures. Interactive systems also have many common-mode elements, units that serve more than one purpose. The rebuilding of a washed-out access road or the need to clear away a toppled construction crane are examples of common-mode failures. Further, as Perrow points out, problems in interactively complex systems are often not revealed clearly. Monitoring is indirect, and causal relationships are not always intuitively obvious. Again, this is true of nuclear plant construction.

Tightly coupled systems, those with little slack or redundancy, are also prone to normal accidents. In these, without a rapid coordinated response, component failure may spread to other units. Unlike an operating nuclear

---

[31] Charles Perrow, "A Framework for the Comparative Analysis of Organizations," *American Sociological Review* 32, 2 (April 1967): 194–208; for a somewhat different interpretation, see Mary Beth Raum, "Decision Anatomies of Three Technology Based Public Bodies in the State of Washington," PhD diss., University of Washington, 1992, 271–282. Senate Inquiry, *Causes of Cost Overruns*, I:35.

[32] Charles Perrow, *Normal Accidents* (New York: Basic Books, 1984), 5.

plant, a construction site is not inherently tightly coupled. Incidents may spread beyond their initial locale or subsystem, but they will probably cause delay, not disaster. "Innocent bystanders" or future generations are not at substantial physical risk. However, delay could gravely wound the organization and the projects themselves. Especially as public and investor scrutiny focused on the Supply System, construction problems magnified political and financial woes. To cite some examples, in 1980, WPPSS was coming to the bond market approximately every forty-five days to borrow $180 to $200 million. For its May bond issue, the Supply System had to pay a hefty 9.5 percent interest rate on its tax-exempt bonds; by September, investors demanded 10.68 percent.[33] On May 18, Mount St. Helens erupted, and the winds carried its greasy ash across the Pacific Northwest. Another discharge a week later spread more ash. Cleanup at the Hanford site of WNP-1, 2, and 4 caused a week's setback in the construction schedule. Meanwhile, at Satsop in western Washington, where the twinned Projects 3 and 5 were going up, a huge derrick, reaching 495 feet above the site base, collapsed on May 29, perhaps due to the volcano's impact. There were no serious injuries, but sections of the derrick fell into the auxiliary building for Plant 3. Since the same crane was slated for later use at Hanford, the accident slowed work on all five projects. In June, the Nuclear Regulatory Commission issued a devastating safety report on WNP-2; work on safety-related systems there effectively ceased for a year.[34] Crippling strikes at Hanford dragged on for months and cost over $700 million. The State Senate Committee's investigation of cost overruns and scheduling delays got underway that summer. Legislators and their staff looked upon the work shutdown with suspicion, sharpening their criticisms of the Supply System.[35] Other forces, both inside and outside WPPSS, were at work as well, but in this instance construction problems were tightly coupled to financial difficulties and a worsening political environment.

Another way to conceive of the pressures facing large-scale project management is to consider the tradeoffs of cost, time and quality. Each goal can be met, in most circumstances, only by sacrificing one or both of the others. In building a luxury home, for example, the cost constraint may be relaxed, but the client may demand the highest quality and insist

[33] See, for example, Joel Connelly, "Record Rate on WPPSS Bonds," *Seattle Post-Intelligencer*, September 24, 1980.
[34] Minutes, WPPSS Board of Directors Special Meeting, January 15, 1982, pp. 6–7.
[35] Senate Inquiry, *Causes of Cost Overruns*, I:45–46.

on speedy completion. On other projects, however, a tight budget and a firm deadline may necessitate cutting back on finish work or eliminating features that are desirable but not essential. At best, project managers should be able to control two of the three variables at the expense of the third. In the case of nuclear plant construction in the 1970s, however, it sometimes seemed that pursuing any one objective necessitated sacrificing the other two and might even prove self-defeating. Indeed, project managers might well feel that efforts to advance even one of the goals would be unsuccessful. For example, since Nuclear Regulatory Commission specifications changed frequently, especially after the Three Mile Island accident in March 1979, attempts to speed up construction were likely not only to cost money but to compromise adherence to standards. Fast-track construction had much the same effect. Work already completed might have to be removed and redone to meet new specifications, costing both time and money that fast track was meant to save. Economy measures were likely to take additional time and mean failure to meet specifications. In an inflationary environment, the costs of delay would be high, and violating regulatory standards could bring about expensive rework orders and, in some cases, heavy penalties. Thus, cost savings would evaporate. A sense of being boxed in pervades the accounts which managers offered as the projects unraveled in the early 1980s. Every effort to cope with one set of problems could exacerbate others.

In the region, Supply System managers were clearly under great pressure to build the projects quickly. They also shared a broader sense that fast-track construction would accomplish the goal. Disparaging the "conventional method of planning, programming, design and construction," one fast-track advocate observed that "in the atomic and nuclear age in which we find ourselves there simply must be a better procedure."[36] It is almost as if its devotees believed the label "fast-track" would in itself ensure rapid completion. In hindsight, this error is obvious. As project management experts Morris and Hough bluntly put it, "Research has shown that on high technology projects, concurrency [overlapping design and production schedules] inevitably leads to project overruns."[37] Indeed, as early as 1976, management consultants advised the Supply System that

---

[36] R. Harvey Self, "Project Management in Construction: Fast-Tracking," in John R. Adams and Nicki S. Kirchof, eds., *A Decade of Project Management* (Drexel Hill, PA: Project Management Institute, 1981), 312. Originally published in *Project Management Quarterly* 5, 2 (1974): 22–24.

[37] Morris and Hough, *Anatomy of Major Projects*, 228–229.

"under fixed-price ... contracting, at least 60 to 70 percent of the engineering design should be completed before construction begins."[38] Such cautions went unheeded. According to conventional wisdom, Northwesterners needed more and more electricity. The region's growth depended on it. Fast-track construction would provide it.

How WPPSS Failed

Speaking in 1992, a decade after he had reluctantly terminated WPPSS Nuclear Projects 4 and 5 and put Projects 1 and 3 into mothballs, Robert Ferguson reflected on the nature of the system's construction problems. "The projects were structured for management ... [where] everything is successful and everyone agrees." (Flyvbjerg and his colleagues describe this kind of assumption as "the 'Everything-Goes-According-to-Plan' type of deception." Indeed, the World Bank has made this into an acronym – the "EGAP-principle."[39]) Ferguson continued, "[V]ery little provision was made for disagreement, for how to handle things when there was disagreement and conflict. And the Supply System obviously was right in the center of ... conflict."[40]

Nuclear construction is inherently extraordinarily complex. (One telling figure comes from a 1979 consultant's report. To build a 1,100-megawatt nuclear plant required thirty-eight million pieces of paper.[41]) The Supply System's organization of construction (Table 4.1) heightened problems of coordination and raised the likelihood of destructive conflict still further. WPPSS employed both of the major types of reactor designs used in the United States, a boiling water model for WNP-2 and pressurized water reactors for the other four plants. Design and engineering work went to three different architect-engineering firms. Three different manufacturers supplied the reactors and the related components that comprise a nuclear steam supply system, the heart of a nuclear plant. To complicate matters further, until 1978 the Supply System left construction management responsibilities in the hands of the architect-engineering firms. That

---

[38] Cresap, McCormick and Paget, *Management Organization and Related Issues*, IV-14–15.
[39] Flyvbjerg, Holm and Buel, "Underestimating Costs ...", 289.
[40] Ferguson interview.
[41] Theodore Barry & Associates, *Management Study of the Roles and Relationships of the Bonneville Power Administration and the Washington Public Power Supply System* (n.p.: January 1979), VII-2. The same study also indicated that the life cycle of a nuclear plant would generate 40 million documents of more than a thousand different types (p. IX-11). Hereafter cited as Barry & Associates, *Roles and Relationships*.

TABLE 4.1. *Construction characteristics*

	WNP-2	WNP-1/4	WNP-3/5
Architect-Engineer	Burns & Roe	United Engineers & Constructors	EBASCO
Nuclear Steam Supply System Source	General Electric	Babcock & Wilcox	Combustion Engineering
Reactor Type	Boiling Water	Pressurized Water	Pressurized Water
Construction Management (after 1980)	WPPSS	Bechtel	EBASCO

Source: Adapted from Washington State Senate Energy and Utilities Committee WPPSS Inquiry, *Causes of Cost Overruns and Schedule Delays on the Five WPPSS Nuclear Plants* (Olympia, WA, 1981), I:10.

year it attempted to centralize by sharing construction management with the A/Es. Two years later, however, WPPSS reversed course and dropped the integrated construction management structure, designating itself as construction manager for Project 2, naming an additional firm to manage the twinned Plants 1 and 4, and returning construction management on Plants 3 and 5 at Satsop to the A/E. Like many who try to account for the failure of the nuclear power dream, Robert Ferguson pointed to the costs of these varied approaches. "Had the Supply System gotten some expert advice in the beginning... like that they should have chosen the best plant and replicated it... They could have just replicated, learned from one, put what you had learned from one into the design, and build not so far ahead."[42]

Failure to standardize reactor design, continually bemoaned by nuclear advocates and indicted by opponents as a sign of the industry's ineptitude, was rooted in the structure of the nuclear industry and the expectations its protagonists shared. Mistaken though it may have been, it was not a matter of willful refusal to take an obviously correct path. The oligopolistic firms manufacturing reactors wanted designs that they could take "off the shelf" when orders were forthcoming. On the other hand, they competed with each other to offer the latest, most advanced reactors and feared the effects of freezing their technologies with a single standard model. In March 1973, the same month that WPPSS received its construction permit for WNP-2, the Atomic Energy Commission introduced new regulatory procedures to pre-approve reactor designs that could then simply

[42] Ibid.

be referenced in utility construction permit applications. It also informed utilities that it would require only a single review process for applications to build duplicate plants. Yet even after these measures, nearly half of the reactor orders failed to take advantage of the streamlining. Industry sources claimed that even pre-approved designs were subject to regulatory changes during the construction process, negating the benefits of standardization. However, as John L. Campbell has shown, institutional problems in the nuclear power industry itself hindered standardization. Manufacturers continued to feel the tensions between standardization and competition. Architect-engineers sometimes lacked enthusiasm for standardization because their profit margins were contractually fixed. Standardization offered little to them. Utilities themselves feared that committing to a pre-approved standardized design might raise costs by locking them into purchasing components from specified suppliers. Both manufacturers and their customers had to worry that incorporating standardized designs from a single manufacturer into a construction permit application would run afoul of antitrust provisions or violate competitive bidding requirements.[43]

That each of the first three WPPSS plants was to be designed by a different architect-engineering firm further complicated the construction challenge, but it was understandable in the context of the times and the region's planning process. When the Supply System took on Projects 1 and 3, utilities around the country were competing to sign up a handful of A/E firms for the 120 nuclear projects they were planning in 1973–74. Although WPPSS hoped to use Burns & Roe, already the A/E on Project 2, for WNP-3, it turned to EBASCO because Burns & Roe was too busy to take on the added contract. In the region's Hydro-Thermal Power Program, the project that became known as WPPSS Nuclear Project 1 (WNP-1) had previously been designated as a redesign of the Hanford N-Reactor's generating plant. United Engineers & Constructors had been hired to do that project, and the firm continued as A/E when the remodeling plan turned into the new WNP-1 project. Similarly, WPPSS found itself with three different nuclear steam supply systems (reactors and related plant elements) for its first three projects. Supply System managers feared that specifying a particular design could run afoul of the state of Washington's statutes mandating competitive bidding on its contracts. Ironically, an organization (and an industry culture) that valued economies of scale

[43] Campbell, *Collapse of an Industry*, 31–49.

and system integration allowed itself to proceed with a hodgepodge of reactors and contractors for its projects.[44]

The Supply System's decision to build pairs of duplicate plants (WNP-1/4 and 3/5) appealed to economy-minded WPPSS leaders in the mid-1970s (see chapter 3). However, the potential savings soon were swallowed up by the general escalation of costs and the complications of trying to build two enormous nuclear plants on the same site. In fact, since each twin at the two sites was being built under different financial arrangements (WNP-1 and WNP-3 net billed, WNP-4 and WNP-5 with direct participant shares, and with 30 percent of WNP-3 and 10 percent of WNP-5 in private utility hands), pairing led to near-endless conflict over allocation of costs at each site. The "cost-sharing" legal cases dragged on well into the 1990s.

A narrative of the construction history of the WPPSS nuclear projects would tell a repetitive story of strikes, weather problems, equipment failures, contractor conflicts, and other problems. In this account, each episode typically begins with bad news, an event that interferes with the swift completion of these urgently needed projects. Then the Supply System responds with an investigation, a meeting, a statement demanding progress, perhaps a management realignment or even legal action. Along with this type of reaction, the Supply System often tries to recount its own version of the episode, placing blame and assessing the impact of the difficulties and promising a better future. Occasionally, questions are posed, challenging the "official" interpretation of the episode, but they are seldom resolved in the Supply System's account. Punctuating the narrative is a series of management studies, each trying to assess how the story has unfolded and where it should move in the future. The Supply System responds to these studies with a mixture of credulity and defensiveness.

It is hard to dispute that the Washington Public Power Supply System, in agreeing to build five large nuclear power plants at a time when nuclear power nationally and internationally faced severe challenges, had taken on enormous tasks. The agency itself lacked many of the capabilities that a larger, more experienced and more sophisticated organization might have brought to its mission. Throughout the years of its most intense activity, the Supply System faced major problems in contracting with hundreds of companies, in arranging the management of construction work forces

---

44 D. Victor Anderson, *Illusions of Power: A History of the Washington Public Power Supply System* (New York: Praeger, 1985), 95–97.

that had reached 8500 by the end of fiscal 1979, and in dealing with the changes in its designs and work plans as the projects progressed.[45]

In early 1979, Projects Director Frank McElwee briefed the Board of Directors on project management issues. McElwee summarized a tangle of interconnected problems:

Almost without exception the contractors work to their own profit objectives rather than to our schedule objectives. The Architect Engineer designs and design reviews are often not timely, and the design budgets are escalating. The multiplicity of changes emphasize [sic] the need for control of change early in the process rather than by contract administration procedures after the design is released for construction. We had 74 labor disturbances in the last eight months of 1978 ... The partial duplication of the non-manual staffs among the contractors, the construction managers and the Supply System have [sic] increased non-manual costs.[46]

McElwee, one of the most candid and insightful of the Supply System's senior managers, had outlined a daunting set of challenges to the massive construction program.

Contracting and Change Orders

The Supply System and its critics both came to believe that there were fundamental problems with the ways in which WPPSS contracted with the companies designing and building its nuclear plants. The laws of the state of Washington posited contracting relationships that were inappropriate for the complexities of nuclear ventures. Contracts of $10,000 and more, according to state statutes, had to be let by competitive bidding, and Supply System management at first interpreted the laws to mean that these had to be lump-sum, fixed-price agreements. Some of these contracts entangled WPPSS in a web of unsustainable commitments. There was virtually no procedure to decide which bidders were qualified to do the work. Contracts would go to the low bidder regardless of the firm's capabilities or the likelihood that it could fulfill the contract terms properly. Several large, experienced construction contractors chose not to go after Supply System business under such conditions. Furthermore, hoping to maintain good relations with local and regional small business interests, WPPSS followed a policy of awarding many small contracts

---

[45] The workforce estimate comes from the Supply System's 1979 *Annual Report*, "The Supply System at a Glance," no page number.

[46] "Presentation on Project Management Improvements to the Board of Directors," attachment to Minutes, WPPSS Board of Directors, January 26, 1979, p. 3.

rather than consolidating them into a few large pacts.[47] By the end of 1977, there were over 400 contracts relating to WNP-2 alone.[48] Fragmentation not surprisingly bred confusion and interference. Fast-track construction exacerbated coordination problems. Construction management tasks belonged to the architect-engineering firms until 1978; these companies were not themselves general contractors and did not directly hire and supervise construction workers. Moreover, the A/Es were not legally agents of the Supply System; WPPSS bore legal responsibility for each contract. Responsibility without line authority was a prescription for trouble.

Fixed-price contracts, at first glance appealingly simple, proved to be a major problem for the Supply System. They appeared to place risk on the contractor, since they would be required to fulfill contract terms even if their costs grew, but in practice these lump sum arrangements were not viable. A 1976 consultant's report stated the difficulty clearly:

Though contracts are let on a fixed-price basis, they cannot be managed as fixed-price contracts in nuclear construction because of the magnitude of changes that inevitably occur ... Traditional fixed-price contracts in effect place all risk on the contractor, which is generally acknowledged to be unreasonable in the nuclear industry because delays are often caused by the owner or by other organizations aside from the contractor.[49]

In 1980 testimony, Frank McElwee explained why fixed-price contracts failed. On these fast-track projects, the progress of engineering and design work had set the pace for construction. Contractors could not proceed until engineering drawings and instructions had been prepared for their task. Second, the "multiple contracts and interfaces among contractors requir[ed] ... extensive coordination," and this imposed additional costs not contemplated at the time the contracts had been bid. Third, the evolution of project design had caused interferences and rework. The most desirable contractors, "the ones that you really want to bid your work, were not bidding it." In sum, "We simply had to move away from the fixed price contract."[50] Attempting to pin the risk on the contractors had backfired.

---

47 Gleckman, "WPPSS: From Dream to Default," 168.
48 Barry & Associates, *Roles and Relationships*, p. VII-1.
49 Cresap, McCormick and Paget, *Management Organization and Related Issues*, IV-16.
50 Archives of Washington State Senate Energy and Utilities Committee Inquiry, Transcript of Testimony, September 18, 1980, p. 26, in Box 32, Washington State Archives, Olympia, WA. Hereafter cited as Senate Inquiry Archives.

The Supply System then moved toward target man-hour contracts, a common practice in heavy construction. Most of its contracts between 1977 and 1979 contained a target figure for the total number of hours of labor (called "man-hours" in that decidedly non-feminist environment) needed for completing the tasks. A sliding scale of incentives, ranging from 9 to 15 percent of the contract value, rewarded contractors who finished their work on target or faster. If additional labor was needed, the incentive could be reduced to a minimum of 3 percent. Earlier lump-sum contracts were also renegotiated onto a target man-hour basis. However, the 1979 Theodore Barry & Associates study for the BPA pointed out that industry experience had shown "that target man-hours incentives alone may not significantly improve performance and may also greatly expand administrative support requirements."[51] Architect-engineers had warned WPPSS of the Achilles' heel in target man-hour incentives: Stinting on labor to meet or beat the target could result in schedule delays. McElwee confirmed that management had been aware of the difficulty but had not solved the conundrum: "We've not been able to design a schedule incentive that would be effective," he confessed. If WPPSS had been in a better position to penalize contractors for slipping behind schedule, the problem would have been less severe. But contractors could almost always demonstrate that the Supply System's own management failures had caused the delays. When other contractors had not delivered essential materials or when the architect-engineers' designs for a task had not been completed, contractors pointed out, they could not be expected to complete their work on time.[52]

By 1980, the Supply System was "inexorably" drawn to a contracting form that appeared to abandon any pretense of cost control: "unit-price – level-of-effort – plus fixed fee" agreements. Under these, contractors negotiated a price per unit of labor based on anticipated costs. WPPSS then would pay them for each unit employed as well as a fixed sum above that level. The Supply System bore the added expense if tasks required more work than anticipated, and it pledged a profit on top of this. This, as McElwee conceded, was "awfully close to being a cost reimbursable contract which is illegal" and contrary to the public interest. WPPSS had to take special care to ensure that the fee was fixed, not a percentage of the costs incurred. The Supply System, McElwee told investigators, "gave

[51] Barry & Associates, *Roles and Relationships*, p. VII-3.
[52] Senate Inquiry, *Causes of Cost Overruns*, I:27–28; Senate Inquiry Archives, Transcript of Testimony, September 18, 1980, p. 27, in Box 32.

up a lot when we went to this type of contract, but we also got something very important to us. That's the unquestioned right to control and
direct the work."[53] Yet this right was hollow without a system capable
of managing the contractors effectively. In practice, regardless of the pace
of work, the contractors would get regular monthly payments from the
Supply System to cover their costs and the fixed fee. This removed the contractors' incentive to settle labor disputes, the more so because they knew
that they could expect renegotiated contract terms after a settlement to
increase unit payments and stretch out completion schedules.[54]

The evolution of contracting methods indicates how troublesome these
relationships were to the Supply System. However, these arrangements
should take second rank when we assess the basic causes of the construction debacle. With a better organizational environment for the projects
and in an organization willing and able to reassess its goals in the light
of changing conditions, the Supply System might well have been able to
solve problems by realigning its contracts, but instead the agency vacil-
lated between, on the one hand, a rather frantic optimism that the next
policy change, the next consultant's report, would turn the corner and,
on the other, a style of learned helplessness that saw every problem as
beyond its control. J. A. Hare, the Administrative Auditor appointed at
the behest of the Washington legislature, hinted at this. His "Report on
Method of Contracting," concluded, "[F]actors other than methods of
contracting, per se, are more significant to the success of the projects."[55]
The Supply System's contracting methods no doubt raised the price and
delayed the pace of the construction projects, but reforming contracting
terms alone could not solve the Supply System's enormous problems.

One of the worst aspects of contracting problems came in the change
orders that became a motif of contractor relations. Initial contracts failed
to cover all contingencies, so they needed constant revision and updating.
The Supply System lacked effective control over these changes. Managers
could approve small revisions, but bigger ones needed ratification by the
Board or the Executive Committee. Meetings approved change orders
totaling tens of millions of dollars with little or no discussion and, seemingly, no comprehension of how far off course projects were veering.

[53] Ibid., 38.
[54] Senate Inquiry, *Causes of Cost Overruns*, I:28.
[55] J. A. Hare, Administrative Auditor, "Method of Contracting (Survey S1)", submitted to
the [WPPSS] Executive Committee of the Board of Directors, August 22, 1980, p. 2–1,
in Senate Inquiry Archives, Box 5.

Sometimes the Managing Director would report to the Board and Executive Committee how much business it was to transact during the meeting. For example, "[Managing Director Neil O.] Strand ... reported that the total amount of business which the Executive Committee was to approve at the March 24, 1978 meeting was $156,400,317.06 which compared to $35,100,779.05 of business acted upon at the March 10, 1978 meeting." That December, he announced that the total volume of Executive Committee business for 1978 was $1,707,188,248.59. These observations suggest a perverse pride in the magnitude of the system's problems with change orders.[56]

A series of management studies chastised the Supply System for its administration of change orders. Delays were long and got longer. One 1978 report found the average time between initiation of a change order and receiving instructions to proceed was 2.7 months. By 1980, this had stretched to about eight months. The recommended interval was two weeks. According to another study, the Supply System had no individual in charge of the change order process. It had no consistent procedures for pricing these adjustments and negotiated amounts case by case. "Because the contractor knows that it would be extremely difficult and costly for the owner to engage a different contractor to perform the work called for in a change order," the Supply System ended up the loser in these negotiations. Finally, change orders ate up unreasonable amounts of Board and Executive Committee time. In one Executive Committee meeting that had devoted only three minutes to policy, it had taken nearly an hour and a half to handle resolutions, many of these accepting change orders. Consultants recommended that the Board give the Managing Director authority to approve change orders of less than $250,000 without Board or Executive Committee involvement. It took a full three years before the Executive Committee adopted that proposal. In yielding that task, it retained authority over only 5.4 percent of the change orders, but these represented over 98 percent of the dollar amounts involved.[57]

In 1978, conscious of the commitments it had made on the net-billed projects (WNP-1, 2, and 3), the Bonneville Power Administration launched a study of its relationship with WPPSS. There was growing tension between Bonneville and the Supply System. Historian Gary K. Miller notes that System managers had expected Bonneville to confer with them

---

[56] Minutes, WPPSS Executive Committee, March 24, 1978, p. 2; December 15, 1978, p. 2.

[57] Cresap, McCormick and Paget, *Management Organization and Related Issues*, III-10, IV-15; Minutes, WPPSS Executive Committee, September 14, 1979, pp. 16–17.

on the choice of a consultant, but BPA selected Theodore Barry & Associates without Supply System input.[58] The 1979 Barry report reproached the Supply System for its change order procedures. It pointed out that only one of the three architect-engineer contracts on these plants contained a clause which spelled out responsibility for corrective work. Although the Barry study found signs of progress, granting that the Supply System had speeded up its handling of change orders in recent months and had instituted an automated system of tracking their progress, it criticized the system's incomplete data base and the "nebulous" target dates for completing changes. Terms like "ASAP, immediately, upon approval and when needed" were inadequate. Change orders consistently turned out to be more expensive than foreseen; a sample of one hundred found that the Supply System's order-of-magnitude cost estimates were 53 percent lower than the final price. Consequently, the Barry report found that project costs for each of the net-billed plants were running higher than 90 percent or more of the nuclear plants of the same vintage.[59]

Summing up the situation in early 1980, a Boeing management consultant bluntly told the Supply System that its "current system, practices, and procedures for the management of changes would have difficulty passing a professional audit... The present practices are slow, cumbersome and inefficient." Boeing was then hired to implement a new system, but when Robert Ferguson testified at the State Senate's WPPSS Inquiry that October, he said of the new change order system, "It's too cumbersome; I looked at that and it looked like a nightmare."[60] Considering how long the problem had persisted, the nightmare was a recurrent one.

Regulation and Its Discontents

When Supply System managers attempted to explain problems – the difficulties with change orders and, more generally, the cost escalation and scheduling delays – they almost always pointed to regulation as the main culprit. The Atomic Energy Commission and its 1975 successor, the Nuclear Regulatory Commission, were constantly imposing new requirements and altering specifications, even after granting the plants' construction permits. From the standpoint of WPPSS management, regulators forced them to aim at a constantly moving target. Changing regulatory

---

[58] Miller, *Energy Northwest*, 276–278.
[59] Barry & Associates, *Roles and Relationships*, V-4, V-12, VI-31, VI-32.
[60] Senate Inquiry, *Causes of Cost Overruns*, I:40.

TABLE 4.2. *Causes of cost overruns, 1977–1981*

	Percent of total cost	Amount
Regulatory Requirements	50%	$4.2 billion
Strikes/Schedule extensions	15%	1.3
Inflation/Estimating and Design Refinements	30%	2.5
Nuclear Fuel	4%	0.3
Other authorized costs	1%	0.1
	100%	$8.4 billion

Source: Memo, WPPSS Inquiry Staff to Senate Energy and Utilities Committee 17 July 1980 "Review of the WPPSS Explanations of the Causes of Cost Overruns and Schedule Delays," Washington State Senate Energy and Utilities Committee Inquiry Archive, Box 5.

requirements prevented adoption of a standardized reactor design for the industry, slowed siting and construction permit approval for each project, and then forced the Supply System to repeat tasks that had been successfully completed in order to meet evermore costly NRC specifications. Changing regulations had domino effects. If cable trays had to be reinstalled, cabling might have to be removed and then replaced. New standards for placement of pipehangers would make rework necessary throughout the plumbing system. Meanwhile, other crafts would find their work stations occupied. Materials deliveries might be blocked. All this stemmed from bureaucrats' decisions in Washington, D.C., to impose new regulations while construction was underway. Small wonder that nuclear project costs escalated and that these plants were delayed; meanwhile, WPPSS leaders continually warned, Northwest energy demand was outstripping supply. The region drew ever nearer to the day the lights would go out.

The testimony of Lindy S. Sandlin, the Supply System's Manager of Financial Management Controls, before the Washington State Senate Energy and Utilities Committee's 1980 Inquiry exemplified management's attitude. The session of June 28, 1980 began with a staff presentation of the findings on cost overruns that WPPSS had provided the investigators. The explanatory categories are summarized in Table 4.2.

Asked to justify the Supply System's explanations, Sandlin came out swinging. Following the staff summary, he began, "I would like to compliment the staff on making the presentation and understanding it after we had spent about 800 hours of paid time and about 400 man-hours of unpaid time in collecting it and putting it together." He insisted that the Supply System's emphasis on regulatory problems was "absolutely correct, right and, in fact, conservative."[61] Sandlin stressed ripple effects; if regulatory requirements had caused a variance (deviation from initial plans) which then created a cost increase on another contract, he categorized both expenses as results of regulation. All told, he contended, 99 percent of the cost overruns from 1977 to 1981 resulted from forces beyond the Supply System's control.[62]

The staff had outlined the method that Sandlin and his colleagues had used. Each variance was placed in one of ten classifications. Then, working with the five predetermined explanatory categories in the table above, the dollar amount in each of the ten classifications was allocated to one or more of the five categories. In questioning from both the staff and the senators, the procedures encountered harsh critiques. Since the categories had been set in advance, they excluded other possible explanations of cost and scheduling problems. Furthermore, many of the allocations themselves seemed arbitrary. Why, for example, did all the budget increases dealing with erosion control (a huge problem at Satsop, where annual rainfall is approximately 70 inches) and seismic resistance (a troublesome matter, since an 1872 quake had done damage around Hanford) end up attributed to regulatory requirements?[63] Did this mean that the Supply System would not have spent any additional funds to protect against mudslides and earthquakes had it not been for NRC demands? This seemed unlikely, especially since there had been no changes in the government's regulations on erosion controls. The initial classification of "design evolution" drew scorn from legislators. "So," Senator King Lysen challenged, "design evolution is correcting errors in design." An Inquiry staff member confirmed Lysen's suspicion: "'previous design errors and omissions' is part of the definition."[64]

The questioners also pointed out that these classifications were not part of any regular WPPSS reporting or management system; Sandlin and his staff had created them to meet the Inquiry's request for explanations of overruns and delays. Indeed, until the previous year the paperwork for change orders did not indicate whether any given contract revision had been necessitated by regulatory requirements. Moreover, the current

---

[61] Senate Inquiry Archives, Transcript of Testimony, June 28, 1980, pp. 29, 31, Box 22.

[62] Ibid., 129–130.

[63] Satsop rainfall data: Satsop CT Phase II Amendment Application, Section 2.1–2, http//www.efsec.wa.gov/Satsop/Phase%20II/application/2_1SiteDescription.pdf, p. 4, accessed August 26, 2004. 1872 earthquake: "Largest Earthquake in Washington," http://neic.usgs.gov/neis/eq_depot/usa/1872_12_15.html, accessed August 26, 2004.

[64] Senate Inquiry Archives, Transcript of Testimony, June 28, 1980, pp. 16–17.

schema for classifying the causes of change orders used a different set of categories than Sandlin's study. In the end, Sandlin's claim of mathematical accuracy fell apart: The final percentages "were estimates within, you know – we're dealing in an area where it's plus or minus 30 percent anyway. In all of these numbers, when you start dealing with judgments, you're not dealing in numbers to 14 decimal places, and so we feel we're in the area of plus or minus 30 percent on each one of these."[65]

WPPSS's emphasis on regulation conflicted with the findings of several studies. Shortly before the Three Mile Island accident, the Congressional Budget Office (CBO) had issued a report on the national pattern of licensing and construction delays. It found that NRC regulations and the impact of court decisions and referenda accounted for 18.99 percent of the construction delays among eighty-four reactors being built. Over 70 percent of the tardiness came from within the private sector, with causes ranging from labor problems to financial and managerial weaknesses. According to the CBO, the largest single factor in holding back construction nationally was declining demand for electricity.[66]

Studies of the Supply System's own experience roughly paralleled the Congressional Budget Office findings. United Engineers & Constructors, calculating the share of change orders required by regulation at its site (Projects 1 and 4), attributed only 5 percent to this factor. Management consultants Coopers & Lybrand, in a 1978 report on WNP-2, put the share at 11 percent. The Theodore Barry & Associates management study had attributed only about 8 percent of work delays to regulation. J. A. Hare, the Supply System's Administrative Auditor, concluded, "Changes directed by the Nuclear Regulatory Commission were found to be significant but not necessarily controlling and never more important than the lack of timely engineering and procurement."[67] Inquiry staff members, in an internal memorandum, added that "other estimates of the cost of regulatory impact from responsible officials [were] as low as 5 percent of the cost overruns."[68] At the end of his testimony, Lindy Sandlin implicitly conceded that regulation in itself had not been the greatest curse on the

[65] Ibid., 70, 138–139, 50, 45.

[66] United States Congress, Congressional Budget Office, *Delays in Nuclear Reactor Licensing and Construction: The Possibilities for Reform*, February 1979. The analysis of delay causes is at p. 24. The period after Three Mile Island saw heightened regulatory scrutiny, but it is implausible to attribute the bulk of nuclear project delay to regulation.

[67] Barry & Associates, *Roles and Relationships*, VII-15–16; Senate Inquiry, *Causes of Cost Overruns*, I:45.

[68] "The WPPSS Excuses," in Senate Inquiry Archives, Box 6.

Supply System. All U.S. utilities building nuclear plants faced the same Nuclear Regulatory Commission; all had to handle the same paperwork. The total labor time required to construct a plant was about the same in Washington State as it was in the Southeast, where Duke Power Company seemed capable of finishing its nuclear projects in far less time and at much lower cost. "However," continued Sandlin, "they have some very definite advantages over the Supply System... They have their own engineering force. They have their own labor market. They pay about one half of what we do in the labor market. They do not have numerous contractors on site, and you can go on all the way down. In conclusion, I'm saying that, yes, you're absolutely right."[69] Regulation itself could not account for the woes of WPPSS.

Labor Problems

According to Lindy Sandlin's figures, labor problems deserved the blame for $1.3 billion in cost overruns. Beset with a fractious array of unions ready to take advantage of the scarcity of skilled craftspeople in the nuclear construction industry and the isolation of the Supply System's construction sites, contractors faced workers prone to strike in order to protect jurisdictional bailiwicks and extract higher wages and benefits. Since contractors could generally pass on cost increases to WPPSS, the System bore the brunt of labor relations problems.

This account makes some sense. Crafts unions and their members knew they held a strong hand on the WPPSS projects and were not averse to playing it. As Sandlin prepared his analysis, the Supply System was weathering a strike of several union locals in the Hanford area. In June 1980, Frank McElwee reported to the Board's Executive Committee that "critical path work on the Hanford projects [Plants 1, 2, and 4] had virtually stopped." The strike at Hanford lasted nearly six months at a cost of $707 million.[70]

Labor disputes were nothing new for the agency. Three months after WNP-2 received its AEC construction permit in 1973, pipefitters walked off the job, returned for forty-eight hours and then left again. The next winter they were striking again, but this was a wildcat action, and

---

[69] Senate Inquiry Archives, Transcript of Testimony, p. 131.
[70] Minutes, WPPSS Executive Committee, June 13, 1980, p. 3; Miller, *Energy Northwest*, 307.

Managing Director Jack Stein reported that the dispute would soon be resolved.[71] In 1976, they struck once more; this time the conflict lasted nearly six months. The Supply System could do little, since the dispute was with the Mechanical Contractors' Association, not WPPSS itself. Costs and delays mounted. In September the Board's Executive Committee was told "the cost of trying to work around the Pipefitters is totally out of line." The next month, Managing Director Strand estimated that even if the strike were settled that day, it would take another 90 days for work to get back to normal.[72] A long litany of such conflicts beset the projects, especially at Hanford, but Satsop was also not immune. In one particularly contentious month in 1978, ten separate labor disputes slowed the projects.[73] Strikes – some short, others long, some unauthorized, others union-sanctioned – were usually capable of shutting down construction.

The Supply System itself could do relatively little to avoid labor disputes. The Hanford site, about 200 miles from major metropolitan areas, had been a high-wage haven for construction workers since the Manhattan Project days. The Supply System, like other nuclear-related employers in the Tri-Cities area, drew upon a skilled work force that was accustomed to considering itself indispensable.[74] In the 1980 State Senate Inquiry, one WPPSS consultant cited another client's explanation, no doubt facetious, of ongoing labor force problems in the Pacific Northwest: "He said, 'All those California hippies, when they left California and it was out of vogue, they moved up to the Pacific beautiful Northwest, and they're just as much a bum [sic] now as they were then." State Senator Sam C. Guess, a staunch defender of the Supply System, could not let this slight pass uncorrected. He replied, "They've been there ever since the early '40s. And I know and [sic] they're not hippies, I'll guarantee you."[75]

Here too, contracting procedures contributed to labor difficulties. They removed WPPSS from direct relations with the construction unions. The fixed-price agreements in effect assigned contractors the role of

[71] Minutes, WPPSS Executive Committee Special Meeting, June 26, 1973, p. 8; February 22, 1974, p. 2.
[72] Minutes, WPPSS Executive Committee Special Meeting, September 10, 1976, pp. 3–4; "Summary Report for Executive Committee," attachment to Minutes, WPPSS Executive Committee, Seattle, October 8, 1976.
[73] Minutes, WPPSS Executive Committee Special Meeting, Seattle, December 1, 1978, p. 3.
[74] S.L. Sanger, *Working on the Bomb*, Craig Wollner, ed. (Portland, OR: Continuing Education Press, Portland State University, 1995), 79, 84, 89–90.
[75] Senate Inquiry Archives, Transcript of Testimony, June 6, 1980, p. 128, in Box 21.

negotiating with labor. Even as the contracts evolved away from the lump-sum model and WPPSS heightened its stake in construction management, the Supply System stayed out of direct contact with unions. But, as we have seen, with WPPSS virtually guaranteeing to pay costs plus a profit, contractors would receive a monthly payment whether or not construction was going on. They had little incentive to settle disputes. When the Supply System did respond to labor disputes, it did so almost as a bystander. Their reactive stance showed in May 1979, when sixty-five sheet metal workers walked out at WNP-2 in a dispute with the Waldinger Corporation, the contractor. Two weeks later, steamfitters posted picket lines at the project gates, too. At its next meeting, the Board's Executive Committee voted to initiate lawsuits against striking unions on the grounds that the Supply System was a third party suffering damages.[76]

However, costly as strikes were, the broader issue was the Supply System's low level of labor productivity. The organization fretted often about this and, as we have seen, began to realign major contracts in 1977 to reward contractors who met man-hour targets. This proved a mistake because it provided no incentive to meet schedules or contain non-labor costs, and there is scant evidence that it actually enhanced productivity.

One example of productivity problems cited in the State Senate Inquiry was a contract with the giant Morrison-Knudsen Company for concrete work on the Satsop plants. The firm's $40 million bid, for a target man-hour contract, rested on an estimate of 8.6 man-hours per cubic yard of concrete. WPPSS's own engineering estimate at the time was only half that – 17.2 per cubic yard. By December 1980, the Supply System's pessimism had itself proven to be too sanguine. The estimated production rate was down to 33.4 man-hours per cubic yard, and the cost of the contract had ballooned to $214 million. The completion date had slipped back three years. The Morrison-Knudsen debacle – and others of comparable gravity – could not be blamed entirely on the contractor or the work force. As the Senate Inquiry's report bluntly notes, "WPPSS could not hold M-K to the contract terms because WPPSS failed to meet its own obligations which made it impossible for M-K to perform in accordance with the contract schedule. For example, WPPSS failed to provide materials to M-K in a timely manner, did not provide necessary access to work areas and failed to process numerous engineering changes in a timely

---

[76] Minutes, WPPSS Executive Committee meetings, May 11, 1979, p. 4; May 25, 1979, p. 3; June 8, 1979, pp. 2–4.

manner."[77] In fact, the Supply System's business manager at the Satsop site, Chub Foster, testified that the changes on that contract during the startup period came at a pace of about forty a week, "in effect one an hour."[78]

The Senate Inquiry's investigating staff found some workers ready to expose abuses in the construction process that indicated a system unable to maintain even a good-faith effort to build the plants. Wes Skinner, superintendent of a pipefitting crew, testified that he had been ordered by a WPPSS manager to "put new men on and increase our crew size. At that time, he told me that there was one little problem, that I was going to increase my crew size, but I should not increase production." Skinner blamed this directive on the fact that contractors would profit from increasing the manpower on a job, and he contended that "They [contractors] run the cost of the plants up and blame the labor." He recounted another occasion in which a foreman told him managers had observed him in a group of workers "just standing around. I said, 'Sure, I don't have nothing to do.' He said, 'Well, I know it.' . . . He said, 'The next time anyone comes around, pick up a hammer and hit the table.'"[79] These charges brought forth angry denials from the individuals Skinner had charged with condoning malpractices. The Supply System suggested that Skinner made his accusations because he was angry about being laid off.[80] It is hard to judge the validity of these and similar charges, but it would be surprising if there were no such episodes on construction projects of such gargantuan proportions and complexity.

Among the work force management problems WPPSS had to contend with was astoundingly rapid turnover. At Hanford the problem was "tremendous," though less acute at Satsop. In 1978, according to the Theodore Barry management study, "three or four workers must be hired each six to nine months in order to fill one position."[81] Turnover meant that the construction work force was usually at the beginning of its

[77] Senate Inquiry, *Causes of Cost Overruns*, I:32–33.

[78] Senate Inquiry Archives, Transcript of Testimony, September 19, 1980, p. 84, in Box 32.

[79] Senate Inquiry Archives, Folder, "September 17, 1980 (17, 18 & 19)," in Box 5.

[80] Senate Inquiry Archives, "Preliminary Response of the WPPSS to Statements of Wes Skinner before the Washington State Senate Committee on Energy and Utilities on July 18, 1980," Box 5.

[81] Neil Strand characterized turnover as "tremendous" in Minutes, WPPSS Executive Committee, Seattle, June 9, 1978, p. 3. Barry & Associates, *Roles and Responsibilities*, p. VII-11.

learning curve for the jobs they were on. However, the Barry report judged workers' pace (as measured against appropriate norms of skill and effort) to be generally satisfactory. Where productivity slipped was in the utilization of work forces. In a sampling of contracts for the net-billed projects, Barry found utilization rates averaged only 48 percent of total time. Work was delayed, in other words, more than half the time. On WNP-2, the most active and advanced project, with a total work force of 3,750, the utilization figure was just 40 percent. The largest group of causes of work delays was "Schedule control," defined as "waiting and traveling for materials, supplies, and tools." The consultants did not view this situation with alarm, observing that WPPSS was in the normal range for recent utilization studies in the industry.[82] The Barry report concluded that the Supply System had to take on a greater role in managing the work forces at its construction sites. Although they recommended that day-to-day responsibilities should remain with the contractors, the consultants advised the Supply System to assume more control in several specific areas.

Two years later, in its Inquiry report, the State Senate Energy and Utilities Committee did not find progress. The Senate Committee's study put the blame for poor labor productivity squarely on management's shoulders. "[T]o the extent that low labor productivity has contributed to cost and schedule problems, the underlying cause has been the failure of WPPSS management to provide proper and timely material, scheduling support, proper access or otherwise to adequately support the craftsmen on the job." Significantly, Projects Director Frank McElwee agreed with the critical assessment: "Low productivity is generally our fault, management. Either the material is not available when and where it should be or the engineering is not available when and where it should be, or the equipment or we've gotten interferences or our planning is incomplete or what have you."[83]

Conclusion

McElwee's confession brings us to a core problem in the Supply System's undertaking. The organization's ambitions vastly exceeded its capabilities. A nuclear plant was, as McElwee later put it "an order of magnitude, maybe several orders" more complex than a hydropower project, and five

[82] Ibid., pp. VII-14–VII-17.
[83] Senate Inquiry, *Causes of Cost Overruns*, I:45–46.

plants were another order of magnitude more difficult than one.[84] Fast-track construction methods exacerbated the situation substantially. So too did the awkward contracting and oversight relationships that the Supply System found itself enmeshed in. While WPPSS managers exaggerated the impact of regulatory changes, they no doubt also hindered the projects' progress. Finally, a fractious and poorly administered labor force added further obstacles. To attempt to quantify the relative importance of each of these factors appears futile, as even Lindy Sandlin seemed to concede. The problems fed upon each other. For example, regulatory decrees required change orders, but so too did the decision to use fast-track construction methods. Problems with contractor oversight were both cause and effect of labor strife and poor workforce utilization.

More importantly, almost nobody involved with energy policy questioned the Supply System's vaulting ambitions. A revealing dialogue appears in a transcript of one of the early meetings of the State Senate Energy and Utilities Committee Inquiry. As the group worked out its plan of action, Senator Al Williams commented, "I'm assuming that the intention ... is not to question whether a WPPSS plant should be built in the first place, but simply to look at insuring that it is done as expeditiously and efficiently as possible." Committee Chair Ted Bottiger replied that the Inquiry's work plan presupposed no special position on the wisdom of the projects. For Senator Sam C. Guess, a Spokane conservative, even such agnosticism was foolish. "I believe they're already built," he remarked sarcastically – and prematurely. But Senator King Lysen, a Democratic gadfly and sharp critic of the Supply System, sought to widen the study, arguing that an investigation solely of management competence was too narrow.[85] The discussion trailed off without resolution, and the Inquiry went forth without attention to whether the projects should continue or, indeed, whether they should have been started. In its final report, however, it did point out that "The probabilities that all five of the plants can be simultaneously financed, engineered, constructed, and can produce electricity when needed and at an acceptable cost have never been tested. No such review is planned by WPPSS or any of its participants."[86]

In fact, as late as 1980, willingness to question the purpose of the Supply System's projects and to raise doubts about attempting to complete

[84] McElwee interview.

[85] Senate Inquiry Archives, Transcript of Senate Energy and Utilities Committee meeting, June 6, 1980, pp. 10–11.

[86] Senate Inquiry, *Causes of Cost Overruns*, I:58.

them was in short supply in the energy and utilities community. The organization's descent into the construction morass resulted from, or at least was enabled by, a climate of opinion that took the venture of building five large nuclear plants as a given. Solving the problems of fast-track construction with multiple plant designs, sites and contractors in a heavily regulated environment marked by contentious labor relations would have been difficult at best. An organization with the best managerial and technical talent might have fared better. But WPPSS could not or would not question its own mission. WPPSS misdiagnosed the region's energy problems as a crisis of supply and misprescribed large, capital-intensive generating stations using an uncertain and controversial technology as the solution. By 1980, a reckoning was near.

5

Collapse

A peculiar populism pervades the story of the Washington Public Power Supply System. As we have seen (chapter 1), since early in the century public power in the Pacific Northwest had been more than an issue; it had been a cause. Its adherents considered themselves both proponents of regional development and foes of a system that subjected Northwestern-ers to the domination of callous utility monopolists and avaricious Wall Street financiers. Woody Guthrie's paeans to the Grand Coulee Dam and the Bonneville Power Administration composed during his brief employ-ment with the BPA in 1941 endure as evidence of the sense of mission that infused the movement for public power. Eisenhower's emphasis on private-public "partnership" and other compromises (such as the arrange-ment for sharing electricity from the Hanford N-Reactor between public and investor-owned utilities) notwithstanding, the public power commu-nity retained a degree of insularity and suspicion that the special interests were still eager to thwart the people's desire for more electric power. This was a corollary of their commitment to develop dam sites and, later, ther-mal plants.[1]

Ironically, the Supply System's massive undertakings were converting public power dreams into a populist nightmare. Reporter Howard Gleck-man, who covered WPPSS intensively in the early 1980s, recalled that he used to hear that the projects were the public's "last stand against the...investor owned utilities who were getting fat." Supply System boosters insisted that they wouldn't be "taken in and robbed by the Wall

---

[1] For an important perspective on the modern meanings of populism, see Michael Kazin, *The Populist Persuasion: An American History* (rev. ed.; Ithaca, NY: Cornell University Press, 1998).

Street types." But, Gleckman noted, "Lo and behold, that's exactly what happened to them."[2] Project delays and overruns drew WPPSS into a spiral of ever more desperate measures to finance the projects. The Supply System's demand for funds to pay the escalating costs of construction and interest on the accumulated bond obligations made the agency the largest municipal borrower in the United States by the end of the 1970s. More than one out of every five dollars borrowed for electrical and gas public utilities between 1978 and 1980 went to WPPSS.[3] An assortment of counsel, consultants, brokers, and dealers enabled the Supply System's plunge into deeper reliance on other people's money.

WPPSS's emergence in the municipal bond market could hardly have come at a worse time for the Supply System. The market itself was rapidly growing and changing. State and local government security issues, $18.2 billion in 1970, climbed to $48.5 billion in 1980 and $85.1 billion in 1983. About two-thirds of the 1970 bonds were general obligation issues, backed by the tax receipts of the issuer. The rest were revenue bonds, in which a specific stream of the issuer's anticipated revenue was pledged to repayment. By 1983, however, revenue bonds represented nearly three-quarters of the volume of new municipal issues. This type of obligation had grown tenfold, from $6.1 billion in 1970 to $63.6 billion in 1983. Special districts and authorities, rather than general state and local governments, came to dominate among the borrowers. Their issues rose from $5.6 billion in 1970 to $48.6 billion by 1983.[4]

The growth of revenue bonds in the 1970s and 1980s reflected a blurring of lines between public and private ventures. A wide range of undertakings gained access to funds from the sale of tax-exempt securities. Government agencies and special districts issued bonds to finance college dormitories, student loans, community hospitals, and airports, to name a few of the newer kinds of projects. These and other "nontraditional" uses of tax-exempt bonds accounted for more than half of the money borrowed from 1979 on.[5] From the late 1970s until the practice was

---

[2] Gleckman, interview.

[3] Gleckman, "WPPSS: From Dream to Default," 182.

[4] U.S. Bureau of the Census, *Statistical Abstract of the United States: 1992*, 112th ed. (Washington, DC: U.S. Government Printing Office): Table 458, p. 285. Robert Lamb and Stephen P. Rappaport, *Municipal Bonds*, 2d ed. (New York: McGraw-Hill, 1987), 8–9 contains a table from *The Bond Buyer 1985 Municipal Statbook* with similar but slightly different figures.

[5] U.S. General Accounting Office, "Trends and Changes in the Municipal Bond Market as They Relate to Financing State and Local Public Infrastructure," excerpted in House Subcommittee, *The BPA and WPPSS*, I:379–384.

restricted in the Tax Reform Act of 1986, municipalities and other public agencies also often issued industrial development bonds, in essence
tax-exempt funding that went to subsidize firms locating or expanding
in an issuer's jurisdiction. Pollution control bonds filled another rapidly
expanding category during these years. Corporations could use municipal bonds to finance their pollution abatement investments. Borrowings
in the category of "industrial aid" totaled less than half a billion dollars
in 1976 but had reached $8.3 billion by 1982.[6] Thus, public utilities seek-
ing funds for their projects vied with a broad range of other tax-exempt
borrowing objectives.

It is almost a truism that the buyers of tax-exempt bonds are institutions and individuals in high marginal tax brackets who can benefit from
tax-free interest income despite the lower yields of these securities. (From
the mid-1960s through the late 1970s, the interest rate ratio between tax-
exempts and taxable bonds was about 70% for securities with comparable
levels of risk. Since the differential depends in large measure on income
tax rates, it has declined since the early 1980s as marginal rates on high
incomes have been slashed. By 2000, interest rates on municipal bonds had
actually surpassed the rates for thirty-year U.S. Treasury bonds.)[7] During the years of the Supply System's dalliance with the municipal market,
commercial banks, historically the largest holders of these bonds, held
a decreasing share of municipal bond issues. A second group of institutional investors, property and casualty insurers, maintained a nearly-
constant fraction; and individual investors took a growing proportion of
these tax-exempt investments. Unit investment trusts provided a vehicle
for individuals to gain tax-exempt income from a diversified portfolio of
municipal securities.* From 1976, a tax law change also allowed the formation of municipal bond mutual funds. These mutual funds expanded
exponentially in the 1980s. They held $5.2 billion in net assets in 1980,
$78.6 billion in 1985, and $205.0 billion by 1990.[8]

---

[6] "New Security Issues of State and Local Governments," *Federal Reserve Bulletin* 66, 1
(January 1980): Table 1.47, p. A36; ibid. 70, 1 (January 1984): Table 1.45, p. A34.

[7] For the 1960s and 1970s, see General Accounting Office, "Trends and Changes" in House
Subcommittee, *The BPA and WPPSS*, I:375. Recent data from http://www.seasongood.
com/PDF%20Files/Chart%207.pdf, accessed May 6, 2005.

[8] Lamb and Rappaport, *Municipal Bonds*, 19; A. Michael Lipper, "Tax-Exempt Mutual
and Closed-End Funds," in Robert Lamb, James Leigland and Stephen P. Rappaport,
eds., *The Handbook of Municipal Bonds and Public Finance* (New York: New York
Institute of Finance, 1993), 60.

* A unit investment trust (UIT) invests in a fixed portfolio of securities, often bonds. They
share some characteristics with mutual funds.

The growing scale and scope of tax-exempt borrowing brought problems and uncertainties around the country. More borrowers sought funds
for a wider range of projects. The growth of special-purpose revenue
bonds meant that the risk of a loan was increasingly dependent upon
the uncertain fate of a specialized undertaking whose prospects were not
easy to judge. The growing importance of individual investors and the
shrinking role of commercial banks as municipal bondholders probably
also meant that bond buyers were becoming less capable, on the whole, of
evaluating the quality of the securities they were offered. Unit investment
trusts (UITs) and the emergence of mutual funds in the municipal field
further removed the investors from a transparent view of the project on
which the value and safety of a particular bond depended.

Several developments in the 1970s shook the municipal market. New
York City's near-bankruptcy in 1975 led to the creation of a new financing
body, the Municipal Assistance Corporation, which refinanced billions of
dollars of the city's obligations and issued new ones, secured by earmarked
taxes and other urban revenues. The MAC – appointed, not elected –
gained powerful control over New York's finances and budget priorities.
Although New York's woes reflected the city's unique financial and social
history, it was not alone. In 1978, Cleveland was unable to muster the
tax funds to pay holders of its general obligation notes and defaulted on
them. Other cities, including Boston, Newark, Baltimore, and Detroit,
faced crises but managed to pay their debts. One of the nontraditional
special borrowing agencies, the New York State Urban Development Corporation, defaulted on more than $100 million of its notes in 1975. One
indication of growing awareness of risk in the municipal market is that
the gap between top-rated bonds (Aaa in the Moody's Investors Service
rating scheme) and riskier ones (Baa) grew from 63 basis points (a differential of 0.63%) in 1970 to 116 basis points (1.16%) in 1980. Investors
were demanding higher returns for their less secure bonds.[9]

If the bond market that WPPSS encountered was disquieting in the late
1970s and early 1980s, the broader economic context could be described
as frightening. Prices shot up, with the consumer price index rising 11.3
percent in 1979, 13.5 percent in 1980, and 10.3 percent in 1981. Hoping

[9] Ibid., 251–253, 71; Annmarie Hauck Walsh, *The Public's Business* (paperback ed., Cambridge and London: M. I. T. Press, 1980), 135–140; "Bond and Stock Yields," *Federal
Reserve Bulletin* 62, 1 (January 1976), A28 and "Interest Rates: Money and Capital
Markets," ibid., 69, 1 (January 1989), Table 1.35, A28. A basis point is one-hundredth
of a percentage point of interest.

to choke off inflation, the Federal Reserve System tightened monetary policy sharply in the second half of 1979; short-term interest rates soared. The federal funds rate for overnight loans to banks averaged 13.35 percent in 1980 and a stunning 16.39 percent in 1981.[10] The economy slumped sharply in late 1979 and early 1980, recovering somewhat only to enter into the postwar era's most severe recession in 1981–82.

President Carter's 1977 declaration of a "moral equivalent of war" on the energy crisis did little more than inspire cynics to note that the slogan's acronym was "meow." Prosperity and stable energy prices early in his administration had dulled public concern. The Iranian revolution cut off that nation's exports in December 1978 and brought Ayatollah Ruhollah Khomeini to power two months later. The "second oil crisis" drove energy prices up far faster than the overall Consumer Price Index; they increased at a 25.1 percent rate in 1979 and 30.9 percent in 1980.[11]

As Carter groped for a "sexy and affirmative" response (the phrase came from his top energy adviser), the nuclear remedy Richard Nixon had prescribed earlier in the decade seemed more poisonous than palliative.[12] On March 28, 1979, a pump and a valve at the Three Mile Island nuclear plant near Harrisburg, Pennsylvania, failed. Although the extent of the accident and its implications for nuclear safety are still matters of debate, in the aftermath of the crisis politicians and public alike were in no mood for initiatives to expand nuclear energy. The taking of American hostages at the U.S. embassy in Teheran in November 1979 heightened anxieties further. Crude oil prices almost tripled between January 1979 and June 1980. They leaped again in the fall when war broke out between Iran and Iraq, cutting Iran's exports and almost eliminating Iraq's.[13] Only as the United States and other industrialized nations slid into recession in 1981 did the energy crisis begin to abate; the developed world paid a high price to ease the energy crisis of 1979–81.

The broader shifts in the bond market and the ramifications of international energy geopolitics rarely entered the deliberations of the Washington Public Power Supply System. The Board of Directors and management focused on the System's schedule delays and cost increases. But the

---

[10] U.S. Bureau of the Census, *Statistical Abstract of the United States 1996*, 116th ed., Table No. 801, p. 520.

[11] Ibid., Table No. 746, p. 484.

[12] Daniel Yergin, *The Prize: The Epic Quest for Oil, Money and Power* (paperback ed.; New York: Simon & Schuster, 1992), 693.

[13] Ibid., 699–714.

wider context affected the financial crisis that WPPSS underwent. High interest rates, qualms about nuclear energy following Three Mile Island, and an economic slump that simultaneously dampened energy demand and roused anger at the prospect of soaring electricity prices all can be traced to the national and international developments that surrounded the Supply System's own travails.

Selling WPPSS Bonds

The interest rate a bond must pay is, in part, a measure of the risk investors perceive in it. As early as 1979, the municipals market was registering some anxiety about the bonds the Supply System was issuing. Although interest rates on the net-billed Plants 1, 2, and 3 paralleled costs of similar highly rated bonds from other issuers, investors demanded higher returns on WNP-4 and 5 bonds than on comparable offerings. The bond sale of February 14, 1979, paid a 7.13 percent net interest cost as opposed to 6.33 percent for an index compiled by the trade journal *The Bond Buyer*. Joint Operating Agencies in Massachusetts and North Carolina were also financing nuclear projects, but they paid less than WPPSS; the gap widened during 1979 and 1980.[14] The February 1979 sale revealed some looming problems. With this offering of $175 million, the Supply System had borrowed more than a billion dollars in less than two years for these two projects. Yet with increasing cost estimates, WPPSS anticipated the need for another $3.5 billion to complete the plants – more than the initial estimates of the two plants' entire cost.[15] Moreover, the pace of spending was accelerating. The February bond issue left the Supply System with 11.5 months of cash coverage (the amount of time that the cash on hand would last at the current expenditure rate). The next issue, that August, raised almost as much – $150 million – but provided only six months' worth of cash. WPPSS now expected it would have to raise an added $3.92 billion. As a borrower, the agency was running faster but falling farther behind.[16]

In the weeks before the February 1979 sale, staff and board members toured East Coast financial centers, meeting with investment bankers and

---

[14] SEC, *Staff Report*, 99,233.   [15] Ibid., 90.

[16] Ibid., 92. To put the additional financing predictions in perspective, by May 1981, when the Supply System placed a moratorium on WNP-4 and 5 construction, anticipated financing needs beyond the $2.25 billion already borrowed came to $8.93 billion.

the representatives of Moody's and Standard & Poor's, the two main firms that rated the quality of municipal bonds. Harlan (Hank) Kosmata, Manager for Planning and Analysis, reported to the board that "each agency reaffirmed the ratings on the WNP-4/5 bonds." The System's financial advisor, Don Patterson of investment bankers Blyth Eastman Dillon, praised the staff for its presentations in the East, but, after questioning, he conceded that there were some buyers who would not invest in the plants that lacked net billing protection.[17]

Despite his generally upbeat attitude in public on WNP-4 and 5 bonds, Don Patterson expressed more concern in private communications. In a letter drafted to Supply System administrators, he described investors' perceptions of these securities as "not very good." He noted that most purchasers of these bonds were either mutual funds or "'kinky' investors who are looking for yield and discount bonds." In 1985, quizzed by Securities and Exchange Commission (SEC) staff on what he had meant by "kinky" investors, Patterson maintained that he was unsure, and that he had heard the term from his colleagues at Blyth Eastman Dillon.[18]

These investments, however "kinky" they may have seemed on Wall Street, were receiving little scrutiny there. The rapid growth of the municipal bond market in the 1970s had brought with it only a modest increase in the effort devoted to researching the quality of these credit instruments. One small brokerage house, Michael A. Weisser, Inc., did report in February 1979 that it had "growing concern" about WPPSS and flatly recommended that "all issues...in any portfolio should be sold...Left on its own, it appears the Authority has positioned itself into an ever deepening hole." It repeated this advice in April 1980; sell all WPPSS bonds "regardless of price."[19] Yet most brokerage firms did not know much about the Supply System. Howard Gleckman commented that nuclear power plants weren't easy for investment researchers to analyze. "You couldn't expect [them]...to understand without a lot of work what made a nuclear power plant...All you could do was...go out there and you'd see this huge pile of cement, you'd see these miles and miles of pipes...but you couldn't really tell if this was progress."[20] Accustomed to studying numbers on

[17] Minutes, WPPSS Executive Committee Special Meeting, February 9, 1979, p. 11; Board Special Meeting, February 14, 1979, pp. 9–11.
[18] SEC, *Staff Report*, 100–102.
[19] Michael A. Weisser, Inc., "Municipal Credit Report," February 15, 1979, and April 18, 1980, reprinted in House Subcommittee, *The BPA and WPPSS*, I:187–189.
[20] Gleckman, interview.

printed pages, bond analysts were slow to grasp the Supply System's construction and cost problems.

To overcome vague worries on Wall Street, cosmetic adjustments in financial practices might suffice for the Supply System. Patterson recommended changes in the form of the official statements required for bond issues and a revised method of calculating interest rates, using a method known as Net Interest Costs. The Net Interest Cost method would make it more attractive to offer Supply System securities as "discount" bonds, to be sold at a price below their "par value." More substantially, he and others pressed WPPSS to complement long-term bond issues with short-term borrowings. This culminated in a proposal from Blyth Eastman Paine Webber (the newly merged brokerage house remained financial advisor to the Supply System) in April 1980 urging the agency to adopt a "Balanced Financing Program." The plan called for WPPSS to switch half of its borrowings to intermediate (seven- to twenty-year maturity) bonds and short-term (two- to seven-year) bond anticipation notes.

The Balanced Financing Program's presentation to the Supply System's Board was revealing. The report, liberally sprinkled with simple graphics, catered to the financially unsophisticated. The final page showed a cartoon of an investor at his desk speaking into the telephone. Short-, intermediate-, and long-term WPPSS securities sit in front of him. "I'll take a million of each," he proclaims. Yet the report's introduction signaled worry about the Supply System's finances: "It is becoming increasingly clear that the Washington Public Power Supply System has entered into a difficult period in its long-term debt marketing program."[21]

Although the Board unanimously approved the Balanced Financing Program for Projects 4 and 5, the plan was never implemented. The eighty-eight utilities participating in the plants each needed to approve the arrangement. They would have to promise to repay the short-term debt to make these notes marketable. The brokerage house that advised several of the larger utilities concluded that short-term borrowing might strain utilities' finances when the notes came due; furthermore, they advised, short-term notes were unlikely to reduce interest costs or to make institutional investors more willing to purchase WPPSS securities.[22]

Less than a week after the Board considered the Balanced Financing Program, it faced another ominous sign of the times. The Supply System,

---

[21] Washington Public Power Supply System, *A Balanced Financing Program*, Prepared by Blyth Eastman Paine Webber Incorporated, April 1980, p. 32, Introduction.

[22] SEC, *Staff Report*, 108–112.

under Washington State law, had to seek competitive bids from investment bankers for its bond issues. Until April 1980, all its offerings had attracted bids from two or more brokerage syndicates, but when WPPSS tried to sell a $175 million issue that month, it drew only one bidder, a syndicate headed by Salomon Brothers and Merrill Lynch. Upon Patterson's advice that the interest rate was "somewhat higher than it should have been," the Board rejected the proposal and exercised its legal rights to negotiate a sale subsequent to an offer it had rejected. Negotiations resulted in savings of more than $31 million in future interest payments in comparison with the initial bid. However, the net interest cost was 9.23 percent, nearly a full percentage point higher than bonds sold only five months earlier.

As we have seen (in chapter 4), mid-1980 found the Supply System with Neil Strand as a lame duck Managing Director, a State Senate investigation of the construction morass underway, labor conflict, regulatory sanctions from the Nuclear Regulatory Commission, and even acts of God in the form of Mount St. Helens' ashy deposits. As the bond market began to pay closer attention to its largest municipal borrower, the Supply System and its allies in the regional energy community had to hope for a dramatic change of fortune. But wishful thinking, not drastic action, seemed to be the order of the day. When the Board's Executive Committee met in June and listened in silence to a damaging report on the NRC's criticisms of safety and scheduling difficulties, Bob Murray, Superintendent of Seattle City Light, suggested sarcastically, "Just act nonchalant and nobody will notice."[23]

Budgets and Bonds

If any manager could have pulled the Washington Public Power Supply System away from the precipice, Robert Ferguson was the one. When hired in the summer of 1980, Ferguson was the Supply System's first administrator with broad experience beyond the Northwest public utility community; his local ties, as a political protégé of Senator Henry Jackson, and as former manager of the Hanford Reservation's Fast Flux Test Facility, also looked attractive to the organization.

Ferguson arrived well aware of the problems facing the projects, everything from safety issues and schedule delays to labor problems and "dope and prostitution" at the construction sites. It was "pretty much of a mess."

[23] Minutes, WPPSS Executive Committee, Seattle, June 13, 1980, pp. 7–8; Joel Connelly, "Feds May Stop N-plant Work, WPPSS Told," *Seattle Post-Intelligencer*, June 14, 1980.

Indeed, he was greeted a week after he took the job with a *Wall Street Journal* article informing him that he had "walked into an awesome array of problems" at WPPSS. At the time, he was not questioning the Supply System's objectives: "I had been told that money was not an issue, that basically the problem was to build the plants," he noted in an interview a decade later. He set out to overcome the obstacles that stood in the way of their completion. Labor issues and safety concerns that had slowed work on WNP-2 to a crawl were quickly resolved. "That part of it that I came out to do was pretty well under control" within a few months.[24]

According to Ferguson, people in the Northwest utility industry "seemed to be so caught up with the assumption that whatever happened the projects were needed. Nobody kind of examined that." When he started asking questions "about where's the power going, what's the impact on rates, is demand going to be inelastic...there were just no answers."[25] He recalled that the problems loomed larger after a mid-fiscal year cost review in fall 1980. The cost review clarified the quandary WPPSS was in. The Supply System needed the review to meet its disclosure requirements for the official statements that accompanied bond sales and to fulfill its pledge to Bonneville to deal with problems on the net-billed plants.

The pace of bond sales was accelerating because of construction cost overruns and schedule delays. As the projects slipped further behind, the Supply System had to borrow more to make interest payments on previous bond issues. High interest rates compounded the agency's problems. When the net interest cost on a sale on September 23, 1980, hit 10.69 percent, Don Patterson conceded that "had it not been for the Supply System's critical cash need, he would have recommended the bid be rejected." Indeed, the borrowing provided only three months' worth of cash for WNP-4 and 5. The Supply System used the high cost of borrowing to press its case for state legislation to allow negotiated as well as competitive bond sales.[26]

Participating utilities were growing concerned about the Supply System's ability to complete WNP-4 and 5. At a meeting of the Participants' Committee (an advisory body established under the 1976 Participants'

[24] "Washington Public Power Nuclear Woes Run the Gamut of Industry's Difficulties," *Wall Street Journal*, August 8, 1980; Ferguson, interview.
[25] Ibid.
[26] "WPPSS Rocked by 10.7% Interest Bid on Bond Sale," *Tacoma News-Tribune*, September 24, 1980, cited in SEC, *Staff Report*, 114–115.

Agreements) in October 1980, utility representatives objected to the Balanced Financing Program and even questioned the wisdom of the projects themselves. After several speakers proposed deferring the projects or looking into selling portions of them off, another interjected, "There's another option: that's cancel it." At the same time, Participants worried that publicizing these options would further weaken the Supply System's financial position. As one put it, he wanted "a very clandestine, hurry-up sort of analysis... I'd hate to see us surface them [issues of deferral or termination]. I'd hate to see it come up because of the potential it has for blowing it sky-high."[27] However, when WPPSS quietly began a study of terminating or delaying WNP-4 and 5, the press soon found out that it was underway. The Supply System then awkwardly denied that its study was really important. Ferguson maintained, "I don't think the issue is the scrapping of the two projects. We're looking at a total range of options, it's just prudent management."[28] Indeed, when the study of delaying construction appeared in March 1981, its conclusion that delay would be extremely costly seemed foreordained by the attempts to reassure project advocates that the study was not intended to thwart the undertakings.

The Participants' ambivalence was apparent. While noting the declining demand for power that raised doubts about whether the plants were needed, they also advocated Federal legislation that would allow the BPA to commit itself to marketing the plants' electricity to the entire region. (Proposals for energy legislation for the Pacific Northwest had, since 1977, raised the possibility of a Bonneville arrangement with Plants 4 and 5 similar to the net billing agreements on the first three projects. However, as passed in December 1980, the Pacific Northwest Electric Power Planning and Conservation Act [Pub. L. 96–501, commonly known as the Northwest Power Act] did not require Bonneville to take over or support the financing of Plants 4 and 5.) Five days after the Participants' Committee meeting, utility representatives met Sterling Munro, Bonneville Power Administrator, in a "hard ball negotiating meeting." According to Hank Kosmata, the Supply System's liaison with the Participants' Committee, the utilities who owned shares of the projects' capability "fully were bearing the risk, and... they were getting damn tired of that situation... [T]hey were saying... if you, Bonneville, and you, DSIs, won't come back

[27] Ibid., 149, 153.
[28] Joel Connelly, "WPPSS Plans to Look at Nuclear Plant Cutbacks," *Seattle Post-Intelligencer*, November 8, 1980; "Construction May Stop at Two Nuclear Plants," *Tacoma News-Tribune*, November 8, 1980, cited in SEC, *Staff Report*, 162–163.

into a position of sharing the financial obligations here, we are willing to essentially cut off this. The hell with it."[29]

The utilities had generally operated with little citizen involvement but now were beginning to face constituents who were curious, and increasingly furious, about the rate increases they were facing and the stories the press had been running about the Supply System's problems. In Springfield, Oregon, for example, ratepayers in 1980 were becoming more vocal. When Springfield held hearings on a proposed 23 percent residential rate increase, opponents blamed WPPSS. According to the minutes, "Glenn Sofge...commented that he felt that if the Utility would quit pouring money into the nuclear projects that the rates would be more reasonable." Another ratepayer, representing the activist group Oregon Fair Share, pointedly asked the Board if it realized that nuclear power was far more expensive than hydroelectric energy. A member replied that they knew this, but the Board nevertheless approved the rate increase unanimously.[30] To make matters worse, only two weeks later the Springfield Board found out that BPA was proposing a 50 percent wholesale rate increase for July 1981.

However, the participating utilities and the region's ratepayers were not the ones who controlled the fate of WNP-4 and 5. Through 1979 the bond market had, at a rising price, accommodated the Supply System's construction dreams. The underwriting firms bidding on the WPPSS offerings had formed into two syndicates, one led by Salomon Brothers and Merrill Lynch Pierce Fenner & Smith, the other by Smith Barney, Harris Upham & Co. and Prudential-Bache Securities. Since, by Washington state law, each bond sale required competitive bidding, each group of investment bankers would offer to buy the bonds and the Supply System would accept the bid with the lower interest rate. The Smith Barney and Prudential-Bache group took seven of the nine competitively bid issues through December 1979.

Until the end of the 1970s, the syndicates lent money to WPPSS with what a lay person might consider blasé indifference. Rather than investigate the creditworthiness of the bonds or the viability or necessity of Projects 4 and 5 independently, they accepted the information in WPPSS's official statements virtually at face value. In their defense, when the Securities and Exchange Commission investigated underwriting practices in the mid-1980s, the lenders pointed out that neither law nor industry custom

---

[29] Ibid., 157–158.
[30] Minutes, Springfield Utility Board, June 11, 1980, pp. 4–5.

required them to search behind the issuer's statements in competitive bond sales. Had the sales been negotiated, the underwriting firm managing the offering would help the issuer in devising the official statement and its public finance unit would learn more about the agency and the borrowing. Even on competitive issues, bond dealers had some sources of information. Underwriters on competitive sales took part in consultations with issuers. In the WPPSS case, they went on several familiarization tours of the Supply System's facilities. These produced a few early hints of investor resistance. On a site visit in 1977, representatives of some of the property and casualty insurance companies, prime targets for municipal bond sales, indicated that they did not want WNP-4/5 bonds without "better evidence of participants' ability to pay and/or BPA backing." But Merrill Lynch officials reassured Supply System financial advisor Don Patterson that "price is the cure to saturation." Lowering bond prices raised interest rates, however, so that this prescription raised the Supply System's debt burden.[31] The tours did not remove visitors' anxiety about the projects. As the Securities and Exchange Commission staff dryly noted in its post-mortem on the WNP-4/5 bond sales, "It does not appear that in these meetings the Supply System revealed non-public adverse information."[32] Thus, despite the opportunities to learn about the WPPSS projects, bond dealers kept bidding on limited information. For them, the Participants' pledges to pay come "hell or high water" made close scrutiny of the investments almost superfluous.

Much the same could be said of the agencies – Moody's Investor's Services and Standard & Poor's – that rated the quality of the Supply System's bonds. Both services maintained positive evaluations of the Project 4 and 5 bonds throughout the four years that the bonds were being placed on the market. Moody's rating was A1 and S&P's A+, in both cases investment grade ratings.[33] Like the underwriters, rating agencies employed analysts more accustomed to evaluating financial documents than studying the underlying activities of the issuers. A manager at Standard & Poor's conceded that it was "beyond their scope of expertise" to judge whether the utility Participants actually needed the power from the projects. The senior analyst for Moody's told SEC investigators that his firm was incapable of

---

[31] Gleckman, "WPPSS: From Dream to Default," 179–180.
[32] Minutes, WPPSS Executive Committee, November 2, 1979, pp. 4–5; SEC, *Staff Report*, 173; Gleckman, interview.
[33] In each case, the rating was in the third highest category; the top four classifications are generally considered investment-grade.

evaluating the validity of the bond counsel's statements that the Participants' Agreements were legal.[34] The ratings agencies did not react to the Supply System's mounting difficulties until it was too late.

Nevertheless, by 1979, despite their initial lack of interest in probing behind the facade WPPSS presented, a few brokers were showing increasing skittishness. In February, a Merrill Lynch report on that month's bond issue suggested that cost overruns on WNP-4/5, along with declining power demand forecasts, meant that the bonds deserved only a "conditional low-range 'A'" and that the ratings should be lowered further "unless timely financing and completion of the Projects occur."[35] The analyst who had prepared that report then prepared another critical evaluation on an April 1979 offering for the net-billed plants. For her efforts, the analyst drew the wrath of Don Patterson, financial advisor to the Supply System, who complained to the head of the director of the Merrill Lynch municipal research group. Smith Barney, one of the lead underwriters in the rival bidding syndicate, also expressed concern about the February 1979 bond issue. Their analyst evaluated the bonds for WNP-4 and 5 as "A" quality, a notch below the rating agencies. The underwriters fretted about the torrential flow of WPPSS bonds. In order to remain diversified, institutional investors often limited their stake in any one issuer to a small proportion of their holdings. Some also had requirements that their portfolios not contain more than a certain percentage of lower-grade bonds. Thus, institutions might not be able to keep buying the bonds that the Supply System was putting on the market. When Merrill Lynch managers met in June 1979 to discuss the company's future plans for WPPSS securities, they voiced confidence in the net-billed plants' securities but were "antsy" about 4 and 5. Yet the brokers continued to treat Project 4 and 5 bonds as investment grade. Portfolio saturation and construction problems notwithstanding, the bond analysts could point to the eighty-eight Participants' promises to pay, come hell or high water.[36]

A few others on Wall Street did not like what they saw of WPPSS in the late 1970s. Eliot Greenbaum, a research analyst, told his bosses at several firms about the Supply System's cost overruns and legal peculiarities of Projects 4 and 5, only to meet with laughter and "abuse." When he told acquaintances at American Express that WPPSS bonds were risky, they called his supervisor, saying "Please, we can't listen to him anymore. We love these bonds. We want to continue to buy them." Jeffrey Alexopulos, a

---

[34] SEC, *Staff Report*, 205–206.   [35] Ibid., 178.
[36] Ibid., 179–187.

municipal credit analyst at investment counselor T. Rowe Price, drew simi-lar conclusions about the Supply System but, unlike Greenbaum, managed to persuade his firm to remove WNP-4 and 5 bonds from their mutual fund portfolios. However, because Supply System bonds were paying higher interest rates than others in the same ratings category, most mutual funds and unit investment trusts stayed with them.[37]

When the Supply System rejected the sole syndicate bid for a bond sale for Projects 4 and 5 on April 29, 1980, an unsettling pattern emerged. After declining the bid as underpriced (i.e., the interest rate required was too high), the agency could then negotiate terms with potential underwrit-ers. As WPPSS financial officials talked with underwriters, they learned that these firms were disposing of the bonds they bought by marketing them to individual investors as components of unit investment trusts. One participant's notes indicated that 70 percent of the net-billed bonds were going to institutional investors, but 70 percent of the Project 4 and 5 bonds went to individual investors, either through UITs or directly. In other words, WNP-4 and 5 bonds were going to the least sophisticated category of investors while banks, insurance companies and other large investors shied away.[38]

Following these consultations, Merrill Lynch assumed the role of lead underwriter for a negotiated sale. In negotiated sales of municipal bonds, brokers can be held responsible where the offers are fraudulent according to provisions of the 1933 Securities Act (Section 17a) and the Securi-ties Exchange Act of 1934 (Section 10b-Rule 5). Underwriters use legal counsel to investigate the offering and prepare a letter stating that in the counsel's opinion the issuer's official statement does not violate Rule 10b-5. WPPSS, however, tried to persuade Merrill Lynch not to carry on a full 10b-5 investigation, claiming time pressure and noting that the relevant documents had already been drafted and circulated for the unconsum-mated competitive offering. Merrill Lynch rejected this but agreed that the Supply System's own bond counsel, Wood Dawson Love & Sabatine, could prepare the report. The underwriter's counsel assumed a limited role beside Wood Dawson, and the entire proceedings were completed in less than a week between Merrill Lynch's selection as leader to the formal sale on May 9, 1980.[39]

There were four more bond sales for Projects 4 and 5 as the projects neared their end. The Supply System was pressing the Washington State

[37] House Subcommittee, *The BPA and WPPSS*, I:64–69.
[38] SEC, *Staff Report*, 263.   [39] Ibid., 187–196.

TABLE 5.1. *Bond sales and net interest cost by year, WNP-4 and 5,*
1977–1981

Year	Number of sales	Sales volume	Net interest cost	Bond buyer 20 index	Added interest cost
1977	3	$365	6.07%	5.72%	6.1%
1978	3	470	6.60	5.99	10.2
1979	3	525	7.69	6.65	15.6
1980	4	690	10.47	8.69	20.5
1981	1	200	11.77	9.81	20.9

Source: Adapted from table in SEC, *Staff Report*, 99.
Sales volume in millions of dollars.
Net Interest Cost is simple average of net interest costs for each issue in a year.
Bond Buyer 20 Index is average of the interest costs of an index of twenty general obligation bonds compiled by *The Bond Buyer* at the date of each WNP-4 and 5 bond sale.
Added Interest Cost is the difference between Supply System interest cost and the Bond Buyer 20 Index divided by the Bond Buyer 20 Index. It measures the premium the market demanded to buy bonds for the two Supply System projects.

Legislature for the right to negotiate bond sales, but it put all of these offerings up for competitive bidding. In each case, only one syndicate submitted a bid. Management and advisors conceded that the financing situation was deteriorating, yet WPPSS accepted all these bids.[40] Since even nominally competitive municipal bond offerings did not require underwriters' legal investigations, the brokerage houses never undertook the kind of scrutiny of the Supply System's situation that negotiated sales usually required. The net interest costs on these sales rose throughout 1980, peaking at 12.44 percent for a sale on December 9. This was more than twice the rate WPPSS had paid for borrowings on these projects in 1977 and early 1978. It exceeded the *Bond Buyer* index by more than 200 basis points (2 percent).[41] Table 5.1 traces the Supply System's growing appetite for funds for WNP-4 and 5 and the rising interest rates it had to promise bondholders.

Although Robert Ferguson contended that the impact of construction delay was even more serious than the escalating interest rates, these factors interacted to deepen the Supply System's difficulties. As participating utilities cast a cold eye on the Balanced Financing Program and pressed for consideration of postponing or even terminating the projects, the fiscal

[40] Joel Connelly, "Record Rate on WPPSS Bonds," *Seattle Post-Intelligencer*, September 24, 1980.
[41] SEC, *Staff Report*, 99

mid-year budget estimate presented at internal management meetings on November 16 and 17, 1980, came as a shock: analysts now expected the five plants to cost $20,440,000,000, nearly $4.5 billion more than in the 1981 fiscal year budget published only four months earlier. Ferguson chose to keep the new figure confidential. He doubted the risk analysis methodology behind it. Instead, he asked budget reviewers for an estimate of the direct costs of events such as the strike and the derrick collapse which had not been factored into the 1981 budget figures. The result, a direct impact of $1.379 billion, was the less startling cost increment that Ferguson revealed to the WPPSS Board's Executive Committee on November 21.[42] The Executive Committee was not the only party kept in the dark about the extent of the budgetary problems. Although rumors did seep back to at least one bond analyst at Merrill Lynch, WPPSS did not inform underwriters and bond agencies that analysts were projecting an escalation roughly equal to the initial estimates of costs for all five plants combined.[43]

Despite its efforts to keep the worst budgetary news under wraps, the Supply System faced growing criticism. The State Senate Energy and Utilities Committee's Inquiry issued a scathing report on cost overruns and construction delays in January 1981. WPPSS Board members reacted defensively. "I take issue with their charge of mismanagement. I think there were well-intentioned management decisions...they just did not work as intended," responded Stanton "Nick" Cain, the incoming president of the Board.[44] One vehement defender of the projects, Richland State Senator Ray Isaacson, even linked the need to build the plants to the feminist movement: "If it weren't for all the dishwashers, stoves and dryers, we wouldn't need so much electric power, but that is the price of liberating women for the work force."[45] Still, doubt and even remorse counterbalanced strident efforts to justify policies. Cain's predecessor, Glenn Walkley, had admitted in testimony to the Committee that

[42] Ibid., 75–80. [43] Ibid., 198.

[44] Joel Connelly, "New WPPSS Board President Defends Past Management," *Seattle Post-Intelligencer*, January 24, 1981, A11. Connelly, whose stories had been consistently tough-minded but rather detached in tone, in this article identified WPPSS Board members' occupations. That Glenn Walkley was an 81-year-old sheep rancher, Cain an orchardist, Fisher a retired appliance store owner, and that other members included a muffler shop owner and a retired bank manager was unlikely to enhance readers' confidence in the agency's professionalism.

[45] Joel Connelly, "WPPSS! We Got In Over Our Heads," *Seattle Post-Intelligencer*, November 20, 1980, A1; "Let's Put a Lid on WPPSS Spending, Say Lawmakers," *Seattle Post-Intelligencer*, October 22, 1980.

Board members had been "over our heads" when they decided on reactor designs. "Yes, mistakes were made and that is why we're here," said Ed Fischer, Chairman of the Executive Committee. Replying to accusations about cost overruns, the Supply System's Director of Contracts admitted, "Yes, my mind is boggled sometimes by these things."[46]

Throughout 1980, even critics of WPPSS's construction and cost difficulties generally had assumed that the Northwest would face future energy shortages if the plants were not completed. By early 1981, politicians and utility leaders began to question this axiom. February's announcement of a 50 percent increase in Bonneville's wholesale power rates raised eyebrows. BPA attributed only one-third of the increase to rising costs of the net-billed plants whose costs it underwrote, but the magnitude of the rate rise opened questions about the elasticity of electrical demand. At about the same time, the Pacific Northwest Utilities Conference Committee revised its forecast of regional energy demand sharply downward. In the short run, PNUCC's prediction of 1981–82 requirements was 9.5 percent lower than its estimate only a year previously. Significantly, the reduction of 1981–82 expected demand approximately equaled the amount of electricity that WNP-4 and 5 were slated to produce.[47]

Even Bonneville, which for years had preached the gospel of rapid electricity demand growth and had guided PNUCC's overestimates, was beginning to shift its position. At a meeting of the Participants' Committee in January 1980, BPA Deputy Administrator Ray Foleen warned utilities that their own forecasts were lower than the ones that the Supply System was publishing in its official bond statements. If "we keep publishing numbers and then we ... find out that the forecasts we're using aren't the same forecasts that the utilities are using ... then we got a problem." That October, Bonneville cautiously but clearly indicated that conservation and renewable resources could be alternatives to fossil fuel and nuclear power generation.[48] In March 1981, BPA announced that it would not decide hastily whether to acquire the expected output of Projects 4 and 5. The 1980 Northwest Power Act had prescribed a complicated economic and environmental test for Bonneville's acquisition of resources, but WPPSS proponents had hoped that BPA would deem the plants necessary and, in

---

[46] "It's Too Much, WPPSS Is Told," *Seattle Post-Intelligencer*, March 17, 1981.

[47] SEC, *Staff Report*, 129–131; House Subcommittee, *The BPA and WPPSS*, III:155.

[48] Pope, "Demand Forecasts and Electrical Energy Politics." For a contemporary account of BPA's position, see Joel Connelly, "A Switch: BPA Says Energy Saving Can Work," *Seattle Post-Intelligencer*, October 21, 1980.

effect, regionalize the costs of the plants. However, Bonneville indicated that it might take as long as three years to make a decision.

Peter Johnson, President Reagan's choice as Bonneville Administrator, took control in May 1981 and continued Bonneville's reappraisal. When his experts predicted that long-run demand growth would only be about 1.5 percent per annum, "it was quite a revelation." Johnson later reflected that Bonneville had been in a "period of delusions."[49] Under the administration of Sterling Munro, BPA had criticized WPPSS for failing to fulfill its nuclear goals, but Johnson was questioning those goals themselves. In the mid-1970s, Bonneville had led WPPSS and public utilities into the fourth and fifth plants. Its shift made the Supply System still more vulnerable.[50]

By spring 1981, the Supply System was on a financial treadmill. In the face of high interest rates, it had to accelerate its borrowing program. More and more of its bond proceeds went to pay interest on past borrowings, and thus each new offering provided less and less cash for the construction program. On May 4, it announced further schedule delays on all five plants. The next day it dropped plans for a $200 million bond sale for Projects Four and Five. According to Howard Gleckman, at a series of secret meetings underwriters had told Robert Ferguson that announcing new cost estimates would make it difficult if not impossible to market bonds for the projects that lacked BPA backing. "You have to slow it down," they told him.[51] Eileen Titmuss, a bond analyst for Drexel Burnham Lambert who was probably the most outspoken Wall Street critic, published a report on May 26 setting the probability that WNP-4 and 5 would be abandoned at 50 percent.[52]

Three days later, making a decision he described as "pure hell for me," Robert Ferguson conceded that Projects 4 and 5 had to stop. He called for a year's construction moratorium. The proximate cause of the Managing

---

[49] Peter B. Johnson interview with Gene Tollefson, October 1984. Videotape in possession of Mr. Johnson, lent to author.

[50] Bonneville's trajectory makes an ironic contrast to developments in national energy policy at the start of the Reagan era. The new administration proposed increased funding for breeder reactor development, deregulation of fuel costs, eased licensing procedures for nuclear plants, and deep cuts in conservation research and development. The Department of Energy, though a target for abolition in the early Reagan years, became a largely unabashed booster of the nuclear power industry. Peter Johnson's Bonneville administration stands out as highly exceptional.

[51] Gleckman, "WPPSS: From Dream to Default," 183.

[52] Eileen Titmuss, "Washington Public Power Supply System: Recent Developments," Drexel Burnham Lambert Inc., May 26, 1981, reprinted in *The BPA and WPPSS*, I:295–301.

Director's decision was a new set of budget estimates. If the $20.4 billion projected six months earlier was too disturbing to present to the Supply System Board or include in bond sale official statements, the new number was too shocking to suppress – $23.9 billion for the five plants, $12 billion for WNP-4/5. In less than a year, cost estimates for the projects had leaped eight billion dollars while completion dates receded into the future.[53]

At the time he recommended the moratorium, Ferguson stated publicly that he still believed the region would need Project 4 and 5 electricity and that the plants should and would be completed eventually. He later confirmed that this had indeed been his viewpoint.[54] Board members concurred: "I am convinced the region needs that power," insisted one PUD commissioner. Executive Committee chair Ed Fischer endorsed a delay but castigated outsiders who doubted the projects' worth. "It's easy enough to sit on the sidelines and throw stones, but I have 180,000 electrical customers in this county [Clark County, Washington, across the Columbia River from Portland] and I take the view that I am providing for the future. I feel that the No. 4 and 5 plants will be completed."[55]

Others, however, were dubious. Representative Jim Weaver scorned the delay in typically harsh terms: "It's like freezing rotten fish and hoping after a year that it will be edible; there's no way."[56] Eileen Titmuss issued a new bulletin for Drexel Burnham Lambert customers on June 3 predicting that the projects would be terminated. Presciently, she continued:

If the projects are terminated, then the participants will have to raise rates to cover the costs...Needless to say, this is not going to be terribly popular with rate-payers.

As a result, we foresee the possibility of extended litigation...Should there be protracted litigation, it is possible that there could be an interruption in the flow of revenues securing debt service.[57]

On the same day the press reported Board members' determination to resume the projects, Moody's dropped its rating on WNP-4 and 5 bonds to Baa1, below investment grade. The move "knocks away potential

53 Joel Connelly, "WPPSS boss: Halt plants," *Seattle Post-Intelligencer*, May 30, 1981.
54 Jim Dullenty, "Ferguson asks 1-Year shutdown of 4, 5," *Tri-City Herald*, May 29, 1981; Ferguson, interview.
55 Joel Connelly, "N-plant delay has utilities officials privately worried," *Seattle Post-Intelligencer*, June 11, 1981.
56 Joel Connelly, "4,000 lose jobs at two N-plants," *Seattle Post-Intelligencer*, June 19, 1981.
57 Eileen Titmuss, "Washington Public Power Supply System," Drexel Burnham Lambert Inc., June 3, 1981, reprinted in *The BPA and WPPSS*, I:303.

investors," Titmuss warned. Many unit investment trusts and bond mutual funds had provisos preventing them from purchasing bonds with Baa1 ratings or lower. Moody's downgrading was likely to drive the UITs, which had been purchasing a rising share of WPPSS bonds, out of the market for Projects 4 and 5.[58] Fortunately for the Supply System's morale, Standard & Poor's adjustment was less drastic. They cut the rating from A+ to A, still considered investment grade.

The spillover effect of the WNP-4/5 financial cut off and construction moratorium threatened to be severe. Eileen Titmuss commented dryly in her June 3 report that the net-billed projects were "neither enhanced nor immunized from this new turn of events," even if Bonneville had in effect guaranteed payment on the projects' bonds. As she pointed out, Bonneville's ability to pay utility Participants (in the form of credits on their power bills) for their shares of the net-billed plants' capability depended on the agency's power sales revenue. BPA would have to continue to raise wholesale power rates to meet net billing commitments. Utilities would be pressed to pay and "their ability and willingness to pay remains a source of concern."[59]

Throughout the second half of 1981, Titmuss's scenario seemed ominously close to reality. The Supply System's appetite for funds reached new heights late that summer. It had intended to sell $450 million in bonds for the net-billed plants but decided it needed to raise $750 million in an issue of September 4. The true interest cost reached a staggering 15.24 percent, despite the fact that rating agencies still gave these tax-exempt bonds their highest ratings (AAA from Moody's, Aaa from Standard & Poor's). Bond analyst Howard Sitzer of Merrill Lynch issued a detailed report, "Washington Public Power Supply System: At the Crossroads," casting doubt on the validity of these ratings. Sitzer also presented some calculations that indicated just how heavily the burden of repaying the WNP-4/5 bonds would fall on the fifteen largest participating utilities. In the worst case, that of Franklin County PUD, 1984 power costs could be 92.9 percent higher than 1980 rates. (Ironically, Franklin County is in the Tri-Cities area, adjacent to the Hanford Reservation, a bastion of support for the nuclear vision of the Supply System.[60])

---

[58] Joel Connelly, "Bond rating is slashed for 2 WPPSS N-plants," *Seattle Post-Intelligencer*, June 12, 1981; SEC, *Staff Report*, 219–268 has a detailed description of the role of UITs.

[59] Titmuss, "Washington Public Power Supply System," 303.

[60] Howard Sitzer, "Washington Public Power Supply System: At the Crossroads," Merrill Lynch Pierce Fenner & Smith Inc., Fixed Income Research, July 24, 1981, reprinted in *The BPA and WPPSS*, I:328.

The construction moratorium on WNP-4 and 5 did not allow the Supply System to walk away from these projects, but the stoppage did mean layoffs for 4,000 construction workers at Hanford and Satsop. Some workers interviewed at Satsop seemed unconcerned, "I'll get laid off, but I'll get another job a few days later," commented one ironworker. Some called the halt foolish, but one craftsman told a reporter, "One reactor is enough, we have all that other power . . . The cost here is not justified, there's been too much waste."[61]

The Supply System and the eighty-eight participating utilities could not adopt this construction worker's viewpoint. Their acceptance of the construction moratorium in June 1981 took place in the tense atmosphere of a meeting held in Portland to avoid the requirements of Washington's open meetings law. WPPSS Executive Committee members stayed away from gatherings at the Supply System itself in order to avoid reaching a quorum and triggering the state's requirement for public access.[62] From the standpoint of the Participants, the situation was highly problematic. They already felt pressured by BPA rate increases to cover Bonneville's expenses and interest on the net-billed plants; now they faced the prospect of double-digit or even greater cost increases without getting any power in return. Moreover, WPPSS was demanding Participants' funds to preserve the stalled projects. Ratepayers were restive. Reminding Northwesterners that their electric costs were far below the national average did little to placate groups that, by 1982, were organizing under the title of "Irate Ratepayers." Shrinking demand forecasts made some utilities and their customers even angrier at the Supply System.

In the summer of 1981, then, several of the major participating utilities began to seek ways to avoid bearing the entire cost of the projects, now in mothballs, they had invested in. The Participants' Committee in July called for BPA to purchase the two plants and spread the costs among all of Bonneville's customers. Alternatively, the committee suggested that non-participating utilities and the Direct Service Industries in the Northwest get involved in financing Projects 4 and 5. The call for others to step in fell on unsympathetic ears. "Thank God we decided not to participate in the last two plants," Seattle City Light's Superintendent told reporters.[63]

[61] Connelly, "4,000 lose jobs"; S. L. Sanger, "Satsop workers start moving on," ibid., p. A3.
[62] Joel Connelly, "WPPSS group went to Portland for secret talks," *Seattle Post-Intelligencer*, June 19, 1981.
[63] George Foster and Larry Lange, "Let Seattle share N-cost, say utilities," *Seattle Post-Intelligencer*, July 24, 1981; Joel Connelly, "WPPSS set to pull plug on two N-plants,"

The DSI customers of Bonneville found regionalization no more palatable. From Wall Street, analysts warned that a Bonneville takeover of Projects 4 and 5 would compromise the creditworthiness of BPA itself and threaten the ratings of bonds for the net-billed plants.[64]

Sitzer's report suggested steps to make WNP-4 and 5 bonds salable again on Wall Street. New bond issues would require language giving priority to bondholders for any funds available if the plants were later terminated. Participants would have to fund a reserve account twice the size of the one they already maintained. Finally, utilities would have to raise their electricity rates enough to cover at least 125 percent of anticipated debt service obligations.[65] Such recommendations, made without expectation they would be accepted, not surprisingly failed to please utilities and ratepayers in the Northwest.

Through the 1970s, WPPSS had generally conducted its business with little attention from the centers of political power in Washington state or elsewhere. However, the cost overruns and delays had caught the eye of those who no longer could afford to pass over the stream of bad news emanating from Richland. The day before Howard Sitzer's gloomy reflections appeared, Washington Governor John Spellman and his Oregon counterpart, Victor Atiyeh, appointed a panel of three corporate heavyweights: Edward Carlson, retired President of UAL, the holding company for United Airlines; John Elorriaga, head of one of the region's largest financial institutions, US Bank; and G. H. Weyerhaeuser of the giant wood products company bearing his family name. The governors charged them with examining the alternatives for Projects 4 and 5. The panel abjured any intent to judge whether the Northwest actually needed the electricity the plants might generate and stated, "We approached our task as businessmen, not power experts." They continued, "Questioning the wisdom of [the investment of $2.25 billion in the projects] or the management of its execution will avail us little in charting a wise course for the future."[66] Failing to question either goals or methods doomed the Governors' Panel.

Seattle Post-Intelligencer, October 17, 1981. One reason for Superintendent Joe Recchi's thankfulness was the vehemence of energy activists' opposition to assuming WPPSS obligations. Recchi, said one, would be "hung from a lightpost" if Seattle participated. Joel Connelly, "WPPSS bailout wins approval," *Seattle Post-Intelligencer*, August 28, 1981.

[64] Sitzer, "WPPSS: At the Crossroads," 346.

[65] Ibid., 323–324; Dean Katz, "Merrill Lynch suggests ways to make WPPSS bonds sell," *Seattle Times*, July 24, 1981.

[66] Edward E. Carlson, John Elorriaga, George H. Weyerhaeuser, *Governors' Panel: A Report on the Economic impacts of the Alternatives Facing the Region on Washington Public Power Supply System Units 4 and 5* (Olympia, WA: Governor's Office, 1981), I-2–I-3.

ernors' Panel reached predictable conclusions in their Septem-
eport. The scenarios they traced all produced verdicts that
hould be completed and would be economically valuable. To
accomplish this, they recommended a complicated scheme to apportion
mothballing costs among the Participants, the DSIs, and the region's private utilities. If the Pacific Northwest Electric Power and Conservation
Planning Council, established by the Northwest Power Act, would then
authorize Bonneville to acquire the output of WNP-4 and 5 as part of the
"Federal Base System," the projects could go forward.[67]

The legitimacy problems of the Supply System infected the Governors'
Panel. Its own makeup engendered distrust among those already suspi-
cious that wealthy corporate interests were pulling the strings behind
Northwest energy policy. Furthermore, the panel operated secretively.
It claimed exemption from the open meeting laws of both Oregon and
Washington. No inquirers received a clear explanation of who was paying
its costs. Robert Ferguson also maintained that the material the Supply
System submitted to the Panel was confidential. By the time the report
appeared at the end of the summer, the Panelists' voice was only one in
an agitated cacophony.[68]

How to finance maintaining the stalled projects became the focus of
intense controversy. In August 1981, before the Governors' Panel report
had appeared, the Okanagon County Electric Cooperative, one of the
smallest Participants in the stalled projects, had been the first Washington
utility to reject inclusion in a plan to preserve the units' readiness and prepare to re-enter the bond market. "We've gone about as far as we can go;
I question how utilities can raise rates to pay for these plants if it continues to go on like this," said the Co-op's president. The next night, when
the Participants met to vote on the financing plan, thirteen other utilities joined Okanagon in saying no.[69] The recalcitrant Participants found
they were not alone in their unwillingness to pay into the mothballing
fund. To the surprise of other players, Portland General Electric (PGE)
announced in mid-October that the Governors' Panel plan was too risky
and costly for its shareholders. "They finked out and shirked their responsibility," responded Stan Olsen, a PUD commissioner from Snohomish

---

[67] The Governors' Panel recommendations are summarized in ibid., I-1–I-7.
[68] John Hayes and Donald J. Sorensen, "Governors' panel shrouded in secrecy," *Portland Oregonian*, September 14, 1981.
[69] Joel Connelly, "Tiny utility rejects aid for WPPSS," *Seattle Post-Intelligencer*, August 27, 1981; Connelly, "WPPSS bailout wins approval," August 28, 1981.

County (Washington), the largest participating public utility. Washington Governor Spellman warned that uncontrolled termination "could produce economic catastrophe."[70]

A week later, after arduous negotiations, those who had balked at mothballing plans seemed to be back in the fold. After the Supply System's Board of Directors endorsed the Governors' Panel solution and approved suspending construction on Projects Four and Five until mid-1983, Robert Ferguson immediately flew off to New York to reassure financiers that, in the words of one board member, "WPPSS is not going down the drain." Yet the agreement was shaky. Three members, including Seattle, voted against the plan. The representative of Kittitas County's PUD, the Supply System's smallest member, complained that managers hadn't been forthcoming with information and got four others to join him in support of a resolution calling for a complete financial accounting on the two mothballed plants.[71] Eileen Titmuss issued a report on October 30 enumerating the problems still facing Projects 4 and 5. She noted that WPPSS had briefly suspended payments to contractors on the projects during the frantic final phase of the mothballing negotiations; one could even claim that the Supply System already had defaulted on its obligations. More important, the accord reached in October 1981 carried the projects only to the end of the year. The Direct Service Industries and the private utilities had to reach a further agreement by November 15. If the cost estimates of mothballing proved to be too low, who would put up the extra money? If these impediments brought on termination, she predicted that other problems would ensue. The Participants' Agreements required utilities to begin paying off the debt a year after a termination date had been established. In the interim, WPPSS was to make payments to bondholders. With Projects 4 and 5 out of the bond market, where would the Supply System get these funds?[72]

The Supply System's approval of the mothballing plan coincided with an administrative shift. As the agency had come under greater scrutiny,

---

[70] Joel Connelly, "WPPSS set to pull plug on two N-plants," *Seattle Post-Intelligencer*, October 17, 1981.

[71] Joel Connelly, "Utilities agree on 2 N-plants," *Seattle Post-Intelligencer*, October 23, 1981; Connelly, "Two N-plants mothballed," *Seattle Post-Intelligencer*, October 24, 1981; Connelly, "Roger sparks controversy on WPPSS board," *Seattle Post-Intelligencer*, October 24, 1981.

[72] Eileen Titmuss, "Washington Public Power Supply System – Working Out – An Analysis of the 'Mothballing' Program," Drexel Burnham Lambert, October 30, 1981, reprinted in *The BPA and WPPSS*, I:307–314.

observers had complained that it was too ingrown, its Board of Directors and Executive Committee comprised exclusively of public utility insiders. To make WPPSS more entrepreneurial and responsive to external developments, the Washington State Legislature had passed a measure abolishing the Executive Committee and replacing it with a new Executive Board with four outside members. The law required the outsiders to come from business, finance, or scientific circles or to be "recognized experts in the construction and management of nuclear facilities."[73] Those initially appointed included Edward Carlson, the UAL executive who had served on the Governors' Panel; Charles F. Luce, Bonneville Power Administrator in the 1960s and in 1981 Chairman of New York City's Consolidated Edison; C. Michael Berry, president of Seattle's leading bank (Seattle-First National); and William Roberts, a Portland entrepreneur.

The new Executive Board, however, did little to alter the Supply System's insularity. The overlap with the Governors' Panel indicated that any new blood would enter with a perspective quite similar to the organization's own. Charles Luce, though a dynamic and generally successful utility leader, was no less wedded to supply-side solutions for energy needs than he had been as Bonneville Administrator two decades earlier. Moreover, the outside members found their positions frustrating. The Executive Board had control of construction management, but policy questions – including the advisability of completing WNP-4 and 5 at all – remained in the hands of the Board of Directors. The new arrangement lasted only a few weeks. Roberts resigned for personal reasons in November, and in January 1982 the other three quit with a statement that called the Supply System's legal structure "virtually unworkable."[74]

Growing Opposition

Few could say "I told you so" about the dormant projects with more justification or more fervor than Jim Weaver. The Oregon Democrat had declaimed against the folly of the Supply System's undertakings for years. From 1977 through 1980, Weaver had blustered and filibustered against efforts to pass a regional power bill, fearing that it would bail out the WPPSS plants by having Bonneville assume the obligations of WNP-4/5

---

[73] Minutes, WPPSS Board of Directors Special Meeting, Seattle, May 15, 1981, p. 4.

[74] James Leigland and Robert Lamb, *WPP$$: Who Is to Blame for the WPPSS Disaster* (Boston: Ballinger Publishing Company, 1982), 222–223; Minutes, WPPSS Board of Directors Special Meeting, Seattle, October 15, 1982, p. 2.

and foist the costs on residential and small business customers. He continued his crusade even after the passage of the Northwest Power Act in December 1980, contending that the Act was designed to factor the Supply System's plants into the region's power supply portfolio. Others saw the Act as a compromise between advocates of building more power plants and those who favored conservation, renewables, and measures to modify energy demand. However, Oregonians drew most of their electricity from private utilities; anti-nuclear sentiment focused on PGE and its Trojan Nuclear Plant.

In Washington State, anger over the WPPSS morass followed a different course. Anti-nuclear activists there had divided in 1979–80 between those who wanted to pursue economic arguments against nuclear power and those who stressed the unsolved problems of waste disposal. For a while in 1980, each group had separately pursued petition drives to qualify initiatives for the ballot, finally uniting to put Initiative 383 banning the importation of nuclear waste into the state.[75] Using the slogan "Don't Waste Washington," campaigners for I-383 found themselves with only token opposition. The U.S. Supreme Court in the Northern States Power case of 1972 had affirmed a circuit court ruling that the federal government had almost exclusive regulatory authority over radioactive emissions and wastes. Industry forces could reasonably expect that even if the measure passed it would be overturned in the courts. Thus it came as little surprise that I-383 passed by a three-to-one margin in 1980 – nor that a Federal District Court judge invalidated it the following June.[76]

By spring 1981, however, those who wanted to base opposition to nuclear power on economic grounds came back with a more potent measure. The initiative they drafted, which became I-394, required voter approval for bond issues for utility financing and a cost-effectiveness study for each power project. Chief petitioner Steve Zemke denied that

---

[75] Richard J. Ellis, *Democratic Delusions: The Initiative Process in America* (Lawrence, KS: University Press of Kansas, 2002) reviews and critiques state ballot initiative processes.

[76] Wayne Hideo Sugai, "Mass Insurgency: The Ratepayers' Revolt and the Washington Public Power Supply System Crisis," PhD diss., University of Washington, 1985, 187–275, provides a comprehensive account. Other studies include Daniel Pope, "Antinuclear Activism in the Pacific Northwest: WPPSS and Its Enemies," in Bruce Hevly and John M. Findlay, eds., *The Atomic West* (Seattle: University of Washington Press, 1998), 236–255, and David D. Schmidt, *Citizen Lawmakers: The Ballot Initiative Revolution* (Philadelphia: Temple University Press, 1989), 77–95. An excellent study, Thomas Raymond Wellock, *Critical Masses: Opposition to Nuclear Power in California* (Madison, WI: University of Wisconsin Press, 1998), esp. chapter 3, finds both state regulation and popular insurgency were effective in curtailing nuclear power.

the measure was directed at nuclear energy: "We're not against building nuclear plants, we're just trying to control the spending," he stated. Nearly two decades later, Jim Lazar, another campaign leader, still asserted that he had worked for the initiative "strictly as an economist."[77]

Unlike the preceding year's campaign, both sides threw themselves into the I-394 contest. Backers, under the rubric "Don't Bankrupt Washington," recruited volunteers to staff telephone banks and call supporters of past environmental initiatives, urging them to sign the I-394 petition. Although an exceptionally rainy spring hampered signature collection, petition circulators in shopping centers and on street corners garnered 186,000 signatures by the July deadline. Those fighting restrictions on utility financing called themselves Citizens Against Unfair Taxes and, like the anti-nuclear petitioners, stressed economic arguments.

On election day, I-394 passed solidly, with a 58–42 percent margin. Although many have seen the anti-nuclear movement as a cause that attracted economically comfortable, professional, well-educated upper-middle-class voters, analysis of the initiative vote by counties suggests that hostility to WPPSS was strongest in counties with below-average levels of income and education and higher than average shares of manufacturing employment.[78] The Supply System and its allies had invoked populist themes, but now it appeared that its opponents were better able to tap into that vein of discontent.

WPPSS and its allies took quick steps to nullify the effects of the I-394 vote. Three banks serving as bond trustees for the net-billed Projects 1, 2, and 3 went to U.S. District Court in December 1981. Bonneville joined in the suit, claiming a legal obligation to take part and to pay the banks'

---

[77] "Initiative 394 on energy-project spending off to stormy beginning," *Seattle Times*, July 9, 1981; Jim Lazar, interview by author, telephone, Olympia, WA, August 3, 1992.

[78] A key discussion of the social bases of opposition to nuclear power is Alain Touraine, *Anti-Nuclear Protest: The Opposition to Nuclear Energy in France* (Cambridge: Cambridge University Press, 1983), also his *The Voice and the Eye: An Analysis of Social Movements* (Cambridge: Cambridge University Press, 1981). For other studies of anti-nuclear power protest which generally stress its post-industrial character, see Jerome Price, *The Antinuclear Movement* (Boston: Twayne Publishers, 1982); Dorothy Nelkin and Michael Pollak, *The Atom Besieged: Antinuclear Movements in France and Germany* (paperback ed.; Cambridge, MA and London: The MIT Press, 1982); Dorothy Nelkin, *Nuclear Power and Its Critics: The Cayuga Lake Controversy* (Ithaca, NY, and London: Cornell University Press, 1971). Christian Joppke, *Mobilizing against Nuclear Energy* (Berkeley, Los Angeles, and Oxford: University of California Press, 1993) contends that new social movement theory fits the West German anti-nuclear movement better than American protest. See also Wellock, *Critical Masses*, passim. For more details on the I-394 vote see Pope, "Antinuclear Activism," 247–248.

litigation expenses, despite the complaints of two liberal Washington congressmen.[79] On June 30, 1982, the day before the bond approval requirement of I-394 was scheduled to go into effect, Federal District Judge Jack E. Tanner declared it unconstitutional on the grounds that it impaired existing contracts regarding the bonds issued for the net-billed plants. Early in 1983, a U.S. Court of Appeals ruling upheld Tanner's decision, and the Supreme Court that spring refused Don't Bankrupt Washington's appeal petition, thus closing the case.

In the end, then, the initiative did nothing directly to prevent the Supply System from financing its nuclear visions. Popular anger at WPPSS and the financial interests that attached themselves to it was no match for the judiciary's well-established concern for the protection of contracts. (Two months after I-394 died in the federal courts, however, Washington State's Supreme Court was to take a very different stance on related Supply System issues.)

This is not to say that public opposition to the Supply System's ventures and qualms about nuclear energy were entirely ineffective. WPPSS was operating in an increasingly tense environment by the early eighties. Board meetings, for example, had typically drawn only members, staff and a handful of interested parties during the earlier years. But now meetings became battlegrounds. For example, at a special Executive Board meeting at the Seattle Center in April 1982, about 500 citizens packed the meeting hall while 3,000 supporters of WPPSS rallied outside.[80] Public anger drove at least one longtime Board member to resign. Ed Fischer, a member since 1968, gave up in March 1982 "with mixed emotions of regret and relief." He reported, "I have received numerous anonymous hate letters and some anonymous rotten, foul and profane calls at all hours of the night."[81] However, in the final analysis, Supply System plans collapsed mostly from their own weight.

Termination

Eileen Titmuss's doubts in fall 1981 about completion of Projects 4 and 5 soon proved well founded. The utilities' and DSIs' mothballing pledges

[79] Joel Connelly, "Congressmen seek probe in N-vote battle," *Seattle Post-Intelligencer,* January 6, 1982.

[80] Minutes, WPPSS Executive Board Special Meeting, Seattle, April 28, 1982, p. 18.

[81] Letter, Ed Fischer to Stanton H. Cain, March 5, 1982, Administrative Central File, WPPSS Headquarters, Richland, WA.

in October in effect constituted a bridge loan for a bridge to nowhere. On January 5, 1982, Snohomish County PUD, the largest participant in the plants, approved payments to the preservation fund. However, later that night, Clark County PUD rejected participation in mothballing the reactors. Clark had bought nearly a tenth of the projects' capability, and its disapproval constituted a heavy blow. The small municipal utility of Drain, Oregon, also had refused to pay. It had taken less than 1 percent, but Drain's decision reflected a more widespread concern among the small public utilities participating in the stalled projects. Alan Jones, chairman of the Public Power Council, saw the Clark PUD vote as dooming the mothballing plan.[82] Adding to a sense of crisis was Moody's Investors Service's decision the next day to suspend its ratings on the two plants' bonds and to review its blue-chip evaluation of the net-billed projects' borrowings.[83]

When Supply System Directors gathered in Seattle on January 8, 1982, the end was near. Twenty-seven of the eighty-eight participating utilities had not paid their shares of preservation expenses. Robert Ferguson announced that unless a funding plan was in place within two weeks, he would propose that Projects 4 and 5 be terminated. Charles Luce, at the Executive Board's meeting the same day, called the situation "very discouraging."[84] There was no more optimism at a special Board of Directors meeting a week later. The recently appointed outside members of the Executive Board, noting that the "mothballing program [had run] aground," resigned. Ferguson reported that there was "no prospect" of a funding agreement for preserving the plants and repeated his intention to recommend termination on January 22.[85]

The Board of Directors and Executive Board meetings on January 22 were on one level merely ratifications of the obvious: there was no way to preserve the plants, and thus they had to be terminated. Only one Supply System Director abstained from the termination resolution. Ferguson concentrated on organizational, legal, and financial arrangements for an

[82] Jim Dullenty, "Clark PUD rejects 4, 5 mothball plan," *Tri-City Herald*, January 6, 1982; Joel Connelly, "Utility gives N-mothballing plan a boost," *Seattle Post-Intelligencer*, January 6, 1982.

[83] (Associated Press), "WPPSS 4, 5 credit rating suspended," *Tri-City Herald*, January 7, 1982.

[84] Minutes, WPPSS Board of Directors Special Meeting, Seattle, January 8, 1982, p. 4; Executive Board Special Meeting, Seattle, January 8, 1982, p. 4.

[85] Minutes, WPPSS Board of Directors Special Meeting, Seattle, January 15, 1982, pp. 2, 4–5.

"orderly" termination. Even stopping the projects was not cheap. He estimated total termination costs at $531 million, with $192 million of this anticipated in the next twelve months. The Supply System would ask participating utilities to lend it about half that figure. The previous day, the state legislature had enacted a measure facilitating short-term loans from the utilities to the System.[86]

But if there was consensus on the need for termination, there was plenty of rancor about its causes and consequences. The press reported that managers sat "stone-faced" following the vote while anti-nuclear protesters in the boardroom popped champagne corks to celebrate their victory.[87] The utilities faced much more than the expense of shutting down the construction projects, costly as that would be. Although only 25 and 16 percent completed, respectively, Projects 4 and 5 had already cost $2.25 billion. After June 30, 1983, according to the terms of the "take-or-pay" contracts the eighty-eight Participants had signed, the utilities would have to start repaying the high-interest debt. The bonds for the two terminated plants would require utilities to pay about seven billion dollars over the next three decades. That equaled nearly $7,000 per customer of the Participants. In return, Participants and their ratepayers would receive nothing.

Following passage of the termination resolutions, Ray Foleen introduced the City Attorney of Bonners Ferry, Idaho, one of the minor Participants in the terminated projects. In an angry statement on behalf of Idaho municipal utilities, the attorney warned the Board that the cities would be discussing:

breach of contract, fraud in the inducement, malfeasance in office and tortious interference of contract, among other things. Additionally, we will be examining the validity of the initial action of our cities in entering the original participants contract without a vote of the people. It will be proposed, as well, that legislation be introduced to clarify that our utility departments will qualify for bankruptcy. Our examination will be from a different angle than...in the past. In the past, we looked at things in the light of what was best for the region as a whole...This time...we will examine matters solely from our own interests...[88]

There was no reason to believe that Idaho utilities were the only Participants contemplating drastic remedies for the costs they were facing

[86] Ibid., January 22, 1982, pp. 3–7.
[87] United Press International article, no title or author, January 22, 1982. This and other articles from United Press International (hereafter U.P.I.) were accessed via the Lexis-Nexis Academic database.
[88] Minutes, WPPSS Board of Directors Special Meeting, Seattle, January 22, 1982, pp. 5–6.

on the terminated plants. Other ominous possibilities also emerged at the meeting. Bankers were, Foleen reported, unwilling to lend money for termination purposes. Peter Johnson, Bonneville Administrator, commented that cost-sharing would be a problem. Plants 1 and 4 were twinned on a common site at Hanford, as were 3 and 5 at Satsop. Putting an end to the Participants' plants meant that costs that the terminated twins were formerly sharing would fall on the net-billed survivors. Bonneville was not eager to bear the added costs that termination of Projects 4 and 5 would impose on their net-billed twins. The four private utilities with ownership shares in Project 3 would resist the burden of Project 5's cancellation.

Robert Ferguson noted that he had, throughout the crisis, contended that termination would exacerbate the cost problems of the first three (net-billed) projects. Cost-sharing would turn out to be one of the stickiest issues of the Supply System's crisis. Suits lingered for well over a decade before the complex issues were resolved. Board President Stanton Cain, observing that many Board members worried about personal financial liability in the wake of the termination, brought the group into executive session for nearly an hour to discuss their own potential vulnerability.[89]

Bonneville's rate setting process accentuated the threat that termination posed to low-cost Northwest electricity. Peter Johnson told reporters that the agency would announce a large wholesale rate hike the following Monday; utilities anticipated an 80 percent increase.[90] That it turned out to be only 60 percent proved of little comfort. Cost overruns on the net-billed plants and the need to start paying their bondholders bore the main responsibility for the rate increases.

The region's mood in the winter and spring of 1982 was grim. The national economic recession had hit the Pacific Northwest severely, making massive increases in electric rates especially unwelcome. The growing ranks of Supply System opponents had scored triumphs with the I-394 referendum passage in November and the cancellation of Plants 4 and 5 two months later. But those victories were worth little, except to committed opponents of nuclear power, if termination and cost overruns on the net-billed projects meant soaring costs. The ratepayers' movement that had pushed through I-394 continued to grow. Sue Gould, chair of the Washington State Senate's Energy and Utilities Committee, was alarmed: "It's a very scary situation. These people are very mad, and they're almost uncontrollable." Some local utilities bore the brunt of public hostility. In

---

[89] Ibid., pp. 6–8.
[90] Joel Connelly, "WPPSS kills two N-plants: Giant rate hike looms for utilities," *Seattle Post-Intelligencer*, January 23, 1982.

March, for example, 2,000 protesters marched on the headquart the Snohomish County PUD in Everett, north of Seattle to protest a bond issue for the net-billed projects. Seattle First National Bank, a plaintiff in the lawsuit to overturn I-394, found itself the target of a consumer boycott.[91]

Less than two months after termination, the Washington Energy Research Center, a joint unit of the state's two major public universities (University of Washington and Washington State University) presented the findings of an elaborate study that the legislature had mandated the previous year. The *Independent Review of WNP-4 and WNP-5* had been designed to assess the viability of the projects and to answer a series of eight questions about them. The review's findings emphasized that the Pacific Northwest no longer was facing rapid growth in electricity demand. It forecast 1.5 percent average annual increases between 1980 and 2000, about half the rate of gain that the BPA-influenced PNUCC had predicted the previous summer. In April 1982, Bonneville itself weighed in with a lowered forecast of a 1.7 percent annual rate.[92] Although both prognostications appeared to vindicate the decision to shut down WNP-4 and 5, the Independent Review shrank from endorsing termination and, like the Governor's Panel, urged the Northwest Power Council to allow Bonneville to acquire their output.[93] Utility leaders, business interests and the academics involved in the Independent Review all found it hard to grasp the possibility that termination was a positive step.

Meanwhile, the Supply System also faced massive problems with the three surviving reactors. Paying over 15 percent interest to borrow $750 million for the net-billed projects in September 1981 had been painful. However, at the same meeting in January 1982 that terminated WNP-4 and 5, the Board of Directors discussed borrowing plans for the net-billed projects. When one member "asked if consideration should be given to slowing down the construction of Projects 1 and 3," another replied that new management for these ventures was "making more than satisfactory progress." But the Supply System's hunger for cash was growing, lenders

[91] Victor F. Zonana, "Rebellion Breaks Out in Northwest Over Skyrocketing Electricity Rates," *Wall Street Journal*, March 19, 1982; D. Victor Anderson, *Illusions of Power: A History of the Washington Public Power Supply System* (New York: Praeger, 1985), 151.

[92] Office of Applied Energy Studies, Washington Energy Research Center, Washington State University/University of Washington, *Independent Review of Washington Public Power Supply System Nuclear Plants 4 and 5: Final Report to the Washington State Legislature* (n.p.: Washington Energy Research Center, 1982), 9 and passim; Jay MacDonald, "BPA seeks buyers in the Southwest for expected surplus nuclear energy," U.P.I., April 5, 1982.

[93] *Clearing Up*, No. 29, December 3, 1982, 9.

were demanding higher returns, and each successive serving of bond revenue seemed to carry the net-billed projects a shorter distance toward completion. Additionally, until April 1982, I-394 remained in the courts. If challenges failed, Washington voters would have to approve any bond issue after July 1. The ratepayer uprising that had enacted the initiative in November made passage highly doubtful. Board members deduced from these circumstances that WPPSS needed to get as much money as possible, as fast as possible. Even with anticipated interest rates at 15.5 to 15.75 percent, the Board unanimously resolved to negotiate sale of half a billion dollars of bonds for the net-billed plants. When the Supply System went to market in February, it found borrowers willing to take even more and actually sold $850 million.[94]

Robert Ferguson, the hard-driving executive who seemed to be one person capable of giving the Supply System coherent direction, found himself in the hospital in Seattle at the end of March undergoing coronary bypass surgery. Diet, medication, and exercise had not sufficiently relieved blockage in his arteries. The Managing Director's recuperation was quick, and he was back on the job by early May, but the personal crisis stood metaphorically for the stresses of the Supply System itself.

By late 1981, as we have seen, Peter Johnson was questioning the Supply System's visions of energy growth. Bonneville's staff generally agreed that regional demand increases would not absorb the huge capacity that WPPSS planned to bring on line. Although Bonneville had stood by without direct control as WNP-4 and 5 went through their death throes, the agency had more influence over the net-billed projects. However, slowing down the net-billed projects was not easy. Bonneville was accountable for the plants but the Supply System maintained the authority to build them. Bonneville's own financial woes impelled it to action. With fixed costs accounting for 90 cents of every dollar it spent, demand for the electricity it marketed was down. "We were moving from a period of being slaves to the nuclear plants to [appropriately] being slaves to ratepayers," was Johnson's characterization of Bonneville's transformation.[95]

Johnson leaned heavily on the WPPSS board to slow or halt construction on one of the net-billed plants. With WNP-2 edging toward completion, the choice was between Project 1, at the Hanford Reservation,

---

94 Minutes, WPPSS Board of Directors Special Meeting, Seattle, January 22, 1982, pp. 9–10; "Remaining three hit paydirt in newest bond offering," *Nuclear News*, March 1982, 49.

95 Johnson, Tollefson interview. Johnson argued for managing BPA as if it were a competitive business: Peter T. Johnson, "Why I Race against Phantom Competitors," *Harvard Business Review* 66, 5 (September–October 1988): 106–112.

and Project 3, under construction at the muddy Satsop site. Both projects were well advanced. WNP-1 was 61 percent complete, WNP-3 more than half-finished. Each, however, had a long way to go. The Supply System foresaw the need to borrow about $2.9 billion more to complete the plants in 1986. Another $600 million would have to come from the private utilities for their 30 percent share of WNP-3.

The choice rested more on political than technical considerations. Pro-nuclear sentiment remained strong in the Tri-Cities area near Hanford, still the heart of the military's nuclear manufacturing activities. Although Grays Harbor County, Satsop's home, was a blue-collar depressed area lacking the panache of the Seattle region, western Washington's politics in general were greener than the Tri-Cities'. Moreover, Hanford's WNP-1 was somewhat closer to completion. These factors implied postponement of Project 3. However, the four large investor-owned utilities (IOUs) who owned 30 percent of WNP-3 were not prepared to see their interests compromised by a construction halt. When asked whether they would contest a delay, they responded, "The answer is yes. We will vigorously resist any such efforts... Deferral of construction of WNP No. 3 would be very detrimental to our customers."[96] Additionally, those in the utility industry who believed that eventually all three of the net-billed plants would be needed had another reason for preferring delay on WNP-1. Anti-nuclear sentiment in western Washington and the presence of a large skilled construction labor force with nuclear experience in the Tri-Cities suggested that Project 1, if stopped, would be easier to re-start than the Satsop plant. Ultimately, Peter Johnson, in a letter to Stanton "Nick" Cain, chair of the Executive Board, recommended that WNP-1 be delayed for a period of two to five years. Behind the recommendation was BPA's power to approve or reject construction budgets for the net-billed projects. "I could not," Johnson wrote, "approve a budget presentation or a financing plan inconsistent with this program."[97] For the BPA administrator, an active Republican political appointee in a national administration committed to an aggressively pro-nuclear agenda, traveling to the edge of the Hanford Nuclear Reservation to insist on mothballing WNP-1 was a bold move. Johnson recalled driving there with a police escort as 6,000 workers demonstrated for continuing work. Protesters burned him in effigy. The episode was "a searing experience."[98]

---

[96] Minutes, WPPSS Executive Board Special Meeting, Richland, April 23, 1982, p. 3.
[97] Ibid., p. 4.
[98] Johnson, Tollefson interview; also, Peter Johnson, interview by author, telephone, McCall, ID, January 21, 1993.

Following heated debate, the Executive Board adjourned its April 23 meeting and reconvened five days later in a large auditorium in Seattle. The minutes noted about 500 citizens in attendance and "an estimated 3,000 supporters of the Supply System's projects gathered outside." Although the press estimated the crowd at only 1,200, two and a half hours of criticism from the public made it clear that anger at Bonneville's position was intense. The shutdown would affect almost all of the 6,375 personnel on the WNP-1 site, with employment dropping to 300 within a year. Even opponents of nuclear projects like State Senator King Lysen complained that mothballing Project 3 would have cost less. Pro- and anti-nuclear forces could agree on the charge that Bonneville had exerted its pressure in the interest of the IOU shareholders in Project 3. Predictions through the 1970s had indicated that private utilities in the Northwest would need new power before the publics. Why, asked Supply System critics, should Bonneville give priority to supplying electricity to private utility corporations?[99]

After the public session, the Executive Board reconvened in a meeting room near Sea-Tac Airport. Members showed they, too, were unhappy with the BPA plan. Don Clayhold, representing Benton County, adjacent to Hanford, complained that Johnson had been asked for a way to proceed with all projects but came back with a proposal to delay Project 1. Stan Olsen, of Snohomish PUD, complained that selecting Project 1 for shutdown meant continuing work on the plant that was "the most costly, the furthest from completion and in the 'backyard' of those who most adamantly opposed it." He would vote for the resolution to delay Plant 1, but as a "virtual hostage" to Bonneville's pressure. Despite such distress, Executive Board Resolution 70, providing for a delay of two to five years, passed unanimously. Now there were two ongoing projects, WNP-2 and WNP-3.[100]

From Construction to Litigation

Peter Johnson had stressed to the Supply System that Wall Street would not be receptive to new bond issues if the net-billed projects were not scaled back. Having swallowed the bitter pill of WNP-1's slowdown, in May the organization's leaders decided to gulp a huge draught of new

---

[99] Jay MacDonald, "Pro-nuclear forces make one final push to 'Save Number One'", U.P.I., April 28, 1982.
[100] Minutes, WPPSS Executive Board Special Meeting, Seattle, April 29, 1982, pp. 28–30.

money. Bond sales before July 1, 1982, would not require approval by voters under I-394. WPPSS wanted to beat the deadline and to assure itself of a cash cushion to postpone later confrontation with a distrustful electorate. Originally planning to borrow $590 million, after negotiations with New York brokers, the Supply System added another $90 million for Project 3. That would give it enough cash to continue construction until June 1983. The sale suffered a setback when, on May 17, Moody's announced a downgrading of bonds on the net-billed projects, from AA to A1. For Moody's, the slowdown on Project 1 was less a sign of prudence than of the Supply System's "inability to maintain an already abbreviated construction program...."[101] Fortunately for the borrowers, Standard & Poor's reaffirmed its high rating for the bonds. Regional opponents sought to derail the sale. The anti-nuclear Washington Light Brigade purchased ad space in the *Wall Street Journal* telling potential investors that "Washington ratepayers oppose this sale without voter approval, and are not obligated to repay the debts incurred." The claim that ratepayers would not have to pay back the debts was untrue. The Supply System had beat the deadline that I-394 imposed; eventually, courts ruled that the net billing contracts were legally valid. Yet the warning reflected the perilous status of the Supply System both in the canyons of Wall Street and the forests of the Northwest.[102] Another symbol of the Supply System's travails was a sign protesters displayed at the board meeting that authorized the $680 million sale. Showing a monster devouring sacks of money, the caption read "Enjoy it – it's the last one you're gonna get!"[103] It was.

Following the termination of Projects 4 and 5, Pacific Northwest ratepayer and anti-nuclear activists joked bitterly that WPPSS had promised to supply power without cost. Now, however, it was providing cost without power. Who would bear these costs became the central question facing the institutions and individuals with a stake in regional energy policies. As the Supply System's construction program shrank, legal activity burgeoned. In Springfield, Oregon, Peter DeFazio carried on the fight. In December 1981, he and other ratepayers filed suit against WPPSS and the Springfield Utility Board (SUB) in Lane County Circuit Court. The ratepayers contended that the SUB had lacked the authority to enter into its 1976 Participant's Agreement without voter approval.

---

[101] John McCorry, "Moody's Cuts WPPSS Rating to 'A1'", *The Bond Buyer*, May 18, 1982.

[102] No author or headline, U.P.I., May 18, 1982; the advertisement was to run in the *Wall Street Journal*, May 20, 1982. I have not been able to verify that it actually appeared.

[103] Jay MacDonald, "'Enjoy it – it's the last one you're gonna get!'," U.P.I, May 21, 1982.

In May 1982, Cyrus Noë, an enterprising former restaurant reviewer, began publishing a weekly newsletter devoted to the Supply System crisis and related regional energy concerns. Its title, *Clearing Up*, could refer both to elucidating and to resolving the conflicts and issues surrounding WPPSS and to the need for clearing away the wreckage, physical, financial, and social, of failed projects. Utilities and attorneys paid seventy-five dollars a month for Noë's bulletins and could enjoy the wit and irreverence he brought to potentially turgid material.[104]

In the aftermath of termination, Washington ratepayers tried their own legal actions to invalidate the Participants' Agreements for Projects 4 and 5. However, in the Evergreen State, it was Chemical Bank, official trustee for the bondholders on the canceled plants, that filed the suit which state courts acted on. In May 1982, the New York bank asked Judge H. Joseph Coleman of King County (Seattle) Superior Court for summary judgment that the utilities were obligated to pay bond owners for WNP-4 and 5. That summer, Washington judges ruled that the ratepayer actions covered the same issues as Chemical's suit. The Supply System, although formally a defendant in the Chemical Bank case, agreed with the bank that the 1976 agreements bound the Participants to return the borrowed money. Meanwhile, in Idaho, other alarmed customers challenged the Participants' Agreements under that state's laws.

The cases moved toward trial in an angry, almost desperate, political environment, with ratepayers furious at the prospect of paying for the carcasses of the defunct projects and bondholders aghast at the notion that the "hell or high water" clauses in the Participants' Agreements might not protect them. In the weeks after termination, PUD meetings were packed with protesters. On March 1, for example, 2,500 people filled a gym in the depressed mill town of Shelton, Washington, and cheered Senator Lysen, who told them, "You are not obligated to pay anything if you don't get anything in return."[105]

Recognizing the danger the court cases posed to both bondholders and ratepayers, a variety of interests joined together to look for ways to spread the burden of the abandoned plants while assuring bondholders they would be repaid. One day before an Oregon judge ruled on a key issue in the DeFazio case, the brokerage firm of Shearson/American

---

[104] "Washington public power supply makes money for Mr. Noë," *Wall Street Journal*, April 8, 1983.
[105] "Swelling Ratepayer Revolt Looms Over Plan to Close WPPSS 4, 5," *The Bond Buyer*, March 8, 1982.

Express (holder of $90 million of the 4/5 bonds) announced a refinancing plan. Under the proposal, Bonneville would borrow more than a billion dollars from the Federal Treasury, then lend a comparable amount to the Supply System by buying taxable bonds paying about 7 percent. Investing that money in higher-yield securities, WPPSS could then use its profits to repay and retire the extremely high-yield WNP-4 and 5 bonds. For Shearson/American Express, the plan offered to provide security for the questionable debts the bankers and their customers held, along with the possibility of earning six to nine million dollars in broker's fees for implementing the proposal. For the Supply System and the participating utilities (and their ratepayers) the scheme offered the promise of relief from the crushingly high interest rates that the project bonds bore. To implement it would have required broad support and congressional action. But Northwestern politicians derided Shearson's plan. Not only did Jim Weaver brand it "laughable," but to Washington Democrat Al Swift the financial maneuvers looked like a "shell game." Oregon Senator Mark O. Hatfield and Idaho Representative George Hansen, both Republicans, labeled the plan a federal bailout and indicated they would not support it. Even Northwest utilities were skeptical. A local utility official summed up the distrust succinctly: "This is not a bailout of the Northwest, it is a bailout of the bankers." The Shearson plan, like others before it, died in a matter of weeks.[106]

In pre-trial activity in the DeFazio suit, the parties aligned themselves in complicated ways. While Springfield remained a defendant in the suit and its counsel continued to maintain that the city had possessed the authority to enter into its Participant's Agreement for its share of project capability, utility boards in four small Oregon cities joined the Springfield ratepayers as plaintiffs. Several other Oregon People's Utilities Districts and municipalities intervened as defendants. Yet the array of plaintiffs and defendants did not accurately describe the parties' positions on the validity of the Participants' Agreements. Springfield and some of the other defendant municipalities wanted their authority to sign the Participants' Agreements to be upheld. However, in the spring of 1982, Springfield had itself sued WPPSS, claiming the Supply System had violated the agreements by canceling the plants, and releasing the Participants from their obligations under the pacts. In his newsletter, Cyrus Noë adopted a shorthand reference. DeFazio's suit was Springfield I. The city's action was Springfield II.

The ratepayers raised several challenges in Springfield I. In the first place, the utilities had entered into the Participants' Agreements without explicit authorization in their charters or proper statutory backing. Oregon laws provided ways to cooperate in joint acquisition of electric generating projects, but the utilities had not followed those procedures. In effect, if not formally, by signing the Agreements, the utilities had issued bonds. They had borrowed money and pledged funds from a stream of future revenues to pay off the debt. They could not legally sell such bonds without a vote of their ratepayers.[107] If supporters of the Agreements contended that WPPSS had been the borrower, not the Participants, opponents replied that this constituted an illegal loan of credit from the Participants, backing up another entity's debt with a promise to pay that the utilities had no right to make. Finally, since the Participants' Agreements seemed to compel utilities to raise rates to get revenue to pay the debt, opponents of the contracts maintained that they unlawfully delegated the authority of local officials over utility rates.[108]

The Springfield I case was argued without a jury before Oregon Circuit Court Judge George Woodrich. Even before the trial itself formally opened, Woodrich rendered a set of decisions that constituted a major triumph for ratepayer activists. On September 29, 1982, he ruled that Springfield had lacked the authority to enter into the Participant's Agreement it had signed with very little controversy six years earlier. This was not a project to be paid off through a "special fund," but rather a long-term obligation of the city, one that required a popular vote. The judgment was "a great victory for the ratepayers in the Northwest," proclaimed Congressman Weaver. On the other side, Springfield's attorney Gary McMurry was shocked. "You could...knock me over with a feather," he told reporters. Judge Woodrich scheduled a trial for October 12, but with his summary ruling on the authority question, even McMurry wondered whether there was anything left to be tried.[109]

---

[107] Municipalities and agencies in Oregon and elsewhere could borrow money without voter approval if there were a "special fund" created to repay the obligation. Plaintiffs contended that this special fund doctrine did not apply to the debts created for the WNP-4 and 5 projects.

[108] The Oregon Supreme Court decision of March 20, 1984 (296 Ore. 550) that reversed the trial court's decision in the DeFazio case contains a useful summary of the opponents' arguments.

[109] "Judge throws WPPSS for a loss in fight with utilities," U.P.I. AM Cycle, September 29, 1982; "Ore. Judge Finds WPPSS 4,5 Pacts Violated Law," *The Bond Buyer*, September 30, 1982.

The implications of the Oregon ruling were unsettling but unclear. It pertained to only eleven Participants, Oregon municipal utilities and PUDs holding only eight percent of the project shares. The Agreements contained a "step-up" clause requiring utilities to increase their shares of payments by as much as 25 percent in case other Participants failed to contribute. But would they apply if these utilities had not been authorized to enter into the agreements in the first place? And would other courts now hold that more of the Participants' Agreements were invalid? WPPSS immediately announced it would appeal Judge Woodrich's decisions, but no one knew if appeals courts would reverse him. The legal and political tangles were enough to make almost everyone agree, in the words of the General Manager of the SUB, "It's a horribly complex mess."[110]

The Oregon case vexed investors, brokers, and the Supply System, but within a few weeks they got a boost from the first in a series of rulings in Judge Coleman's court in Seattle. Decisions concerning Washington public utilities were likely to determine whether the Supply System would be able to pay its creditors on time. Chemical Bank, as bond trustee, had sued all Washington Participants as well as the Supply System itself, and on the opening day of oral arguments, almost 100 lawyers filled the courtroom. On October 15, 1982, Coleman ruled that Washington public utilities did have authority to enter into the Participants' Agreements. He left for later resolution several other issues. Were the contracts themselves valid? Had the utilities been coerced into signing them? Were the agreements so unreasonable and unfair to the Participants that they should be voided as "unconscionable"? Had the Supply System breached them by terminating the projects? Chemical's lawyer, Michael Mines, when asked whether bondholders had a good investment, replied cautiously, "There are still issues down the road to be decided."[111]

The autumn of 1982 was a grim time for the Supply System and bondholders who hoped to get payments on the $2.25 billion they had extended for the terminated plants. Utility defendants in the Chemical Bank case maneuvered to delay a trial which might result in an order to start their debt payments as soon as January 25, 1983. In November alone, WPPSS absorbed a series of sharp blows. The elections of November 2 showed that public displeasure with the Supply System had not abated since the passage of I-394 a year earlier. Ten Washington PUD commissioners who served on the WPPSS Board faced votes in their utility districts. Of them,

---

[110] "Ore. Judge Finds...," 30 September 1982.
[111] "Wash. Judge Rules WPPSS 4, 5 Debt Pacts Are Legal," *The Bond Buyer*, October 18, 1982.

five chose not to run for re-election; two others resigned early; one lost in a primary contest, and two in the general election. None returned to office. Anti-WPPSS insurgencies elected eleven of their thirteen candidates.[112]

The day after the election, Idaho's Supreme Court enjoined the small cities of Heyburn and Burley from raising rates to pay for the obligations on their shares of the terminated projects. A minuscule amount of money was involved – Heyburn had only 990 customers, Burley's project share was less than two-thousandths of the total– but once again a court had told Participants who had bought nebulous "project capability" in the terminated plants that they weren't required to pay for the projects. The Idaho cities had faced severe impacts. Burley had recently imposed a 33 percent increase and Heyburn's electric rates had jumped 65 percent, in each case mostly to cover WNP-4 and 5 costs. Notably, although two married couples had brought suit against their municipalities, another plaintiff was the J. R. Simplot Company, Idaho's giant potato processing firm, which had a plant in Heyburn. The ratepayers' revolt manifested anger at financial interests that presumably had been profiting from the Supply System's profligacy, but the grassroots could be fertilized by corporations seeking to maintain their own cheap energy supplies. (The Idaho Supreme Court lifted the injunction three weeks later, ruling that the cities had now shown why they had to raise their electric rates to meet the WPPSS costs. As the ratepayers' suit was proceeding, the cities paid their increased revenues into an escrow account pending a decision on whether they were obliged to pay the Supply System.[113])

Near the end of November, Standard & Poor's lowered the credit ratings for bonds on the terminated plants from BBB+ to B, a level S&P defined as "speculative." The immediate impact seemed small. According to Eileen Titmuss, the market's reaction to this news was "ho-hum." After the Springfield I ruling at the end of September, the bonds' prices had already dropped steadily and had moved sharply down in mid-November, selling for as little as 40 cents on the dollar. The rating cut did more to ratify Wall Street's prior judgment that the bonds were risky than to shift that evaluation.[114]

Three days later, Managing Director Ferguson announced his resignation. Following his heart surgery the past spring, he had returned to work

[112] *Clearing Up*, No. 25, November 5, 1982, 6–7.

[113] U.P.I. Regional News, Idaho, AM Cycle, November 3, 1982 and Nov. 24, 1982.

[114] *Clearing Up*, No. 28, November 26, 1982, 2; Joel Connelly, "Bond rating is slashed for two N-plants," *Seattle Post-Intelligencer*, November 20, 1982.

only five weeks later, promising his doctor it would be on a part-time basis. Soon, though, he was putting six- and seven-day weeks of long hours. On a flight from Washington in mid-November, Ferguson suffered a small seizure and required oxygen. *Seattle Post-Intelligencer* reporter Joel Connelly commented that the hard-driven fifty-year old's "hair is greying, his face is sagging, and walking is an effort for him."[115] The Managing Director could take pride in an accelerated and smoother construction pace on WNP-2, its completion expected in 1983, and continued progress on WNP-3, the remaining Satsop plant. But construction was taking a backseat to litigation and the agency's protracted scramble to find money to pay off its debts and continue its remaining two projects.

Ferguson himself noted that managers in the Supply System had a very high incidence of heart disease. "It is very stressful work," he commented. Worries at WPPSS seemed widespread. An agency that had devoted a substantial amount of time and effort to self-congratulations now showed its anxieties. Asked about whether there was a light at the end of the tunnel, Supply System information officer R. F. Nowakowski replied, "That light could be a locomotive coming at you." Meanwhile, Cyrus Noë examined the initial lists of WNP-4 and 5 bondholders turned over during trial discovery and found "no investor surprises" to report. There was, however, one well-known name on the list – Charles M. Schulz, the creator of the "Peanuts" comic strip. Perhaps Schulz, the man who gave the world the tormented Charlie Brown, was the most fitting investor in the Washington Public Power Supply System as it underwent its own agonies.[116]

---

[115] Joel Connelly, "Stress of the rescue try too much," *Seattle Post-Intelligencer*, December 5, 1982.

[116] Martin Heerwald, "WPPSS' $7 billion question – Who Pays?" U.P.I., Regional News, Idaho, Utah, BC Cycle, November 4, 1982; *Clearing Up*, No. 26, November 12, 1982, 2.

6

Endgame

Ruling on procedural motions in Chemical Bank's effort to force the participating utilities to pay bondholders for their loans to the terminated WNP-4 and 5, the Washington State Supreme Court commented in December 1983 that this was "by any measure, a massive case with enormous stakes."[1] Tens of thousands of bondholders were facing the loss of billions on what had looked like investments with ironclad guarantees of payment. On the other hand, more than a million ratepayers across six Northwest states had to contemplate the prospect of paying an additional $7 billion on their electric bills for projects that would never be completed. Carl Halvorson, a Portland construction executive who chaired the Supply System's Executive Board, told reporters early in 1983 that WPPSS itself was on the brink of "Armageddon."[2] But the WPPSS crisis was more than a simple tug-of-war between implacable lenders and obdurate borrowers. The crisis affected the Northwest's broader economy and polity and had ramifications for the nation's energy policy and municipal finance. First, was the Pacific Northwest's energy future still based on assumptions embedded in Glenn Seaborg's vision of a "Nuplex" of clean, cheap nuclear energy production, the 1968 Hydro-Thermal Power Program, and Donald Hodel's apocalyptic warnings against the "prophets of shortage"? Would the 1980 Northwest Power Act and the Northwest Power Council it established enshrine these supply-oriented policies? Or would awareness of new realities, along with the legislation, make the region a model for the nation in planning that balanced construction and conservation, energy and environment? If the Supply System were

[1] Quoted in *Clearing Up*, December 30, 1982, 1.
[2] Joel Connelly, "WPPSS is nearing 'brink' fast, its chairman warns," *Seattle Post-Intelligencer*, January 21, 1983.

forced to default on its WNP-4 and 5 obligations, would financiers treat the Pacific Northwest like a pariah for future government borrowing? Nationally, would a $2.25 billion municipal bond default (the largest in American history, by an organization that had, until two years earlier, been borrowing more than any other state or local entity in the country) shake the foundations of municipal finance? Did the WPPSS fiasco sound the death knell for the dreams of nuclear power as a cornerstone of a reconstructed American energy supply?

By the end of 1982, the American economy was reviving from a painful recession. Nationally, unemployment, which peaked at 10.8 percent in November and December, began to decrease with the new year. Gross domestic product, after declining in 1982, grew at a 5.0 percent annual rate in the first quarter of 1983 and leaped ahead at a 9.3 percent pace in the next three months. But the pace of recovery was uneven. The Pacific Northwest, its wood products manufacturing base suffering from high interest rates and the housing market's doldrums, lagged behind. Oregon and Washington both had unemployment rates of 13.8 percent in January 1983. They improved only slightly during 1983.[3] The region's slow recovery reflected both the severity of the recession and a secular shift away from resource-extraction and manufacturing. In the state of Washington, for example, manufacturing employment did not return to the peak levels of late 1979 until late in 1987.[4] Boeing and Weyerhaueser had been the keystone firms in the Northwest, but by the 1990s Microsoft, Starbucks Coffee, and Amazon.com would become the pacesetters.

Stagnation and rapid inflation in the 1970s and early 1980s had taught some, if not all, energy planners that the demand for electricity was elastic with respect to both income and price. Even those who had not accepted the new gospel of conservation and renewable energy sources had become more aware that the growth curve of energy consumption had flattened. In April 1983, the Northwest Power Planning Council set forth its first *Northwest Conservation and Electric Power Plan.* Adopting many of the

---

[3] National unemployment figures available at U.S. Department of Labor, Bureau of Labor Statistics website, http://www.bls.gov/cps/cpsatabs.htm, Table A-1: Employment Status of the Civilian Population by Sex and Age. GDP data at U.S. Department of Commerce, Bureau of Economic Analysis, http://www.bea.doc.gov/bea/dn/home/gdp.htm, Interactive NIPA Tables, Table 1.1.1. Percent Change From Preceding Period in Real Gross Domestic Product. Both sites accessed May 25, 2005. For state figures, see U.S. Department of Labor, Bureau of Labor Statistics, *Employment and Earnings* 31, 3 (March 1984): Table D-1, 114–115.

[4] U.S. Bureau of Labor Statistics, *Employment, Hours and Earnings: States and Areas, 1972–1987*, Bulletin 2320, March 1989, Volume V, Puerto Rico-Wyoming, 3473.

methods and assumptions of those who had been labeled heterodox a few years earlier, the Council offered not a single prediction but a range of growth rates, based on varying assumptions about recovery from the recession, prices and employment growth. The "medium-low" forecast was for 1.5 percent yearly from 1981 to 2002. The "medium-high" figure was 2.1 percent. Based on these forecasts, the Council set forth a "resource portfolio" of energy supplies for the Northwest to maintain or acquire. It was this portfolio that effectively doomed any chance of restarting Projects 4 and 5. The plan pointed out that even under the most rapid growth scenario (considered "highly unlikely"), no new large generating plants would be needed in the next fifteen years. The Pacific Northwest, contrary to all previous conventional wisdom, had a surplus of electricity, not a shortage.[5]

Even before the Power Council rendered its verdict on Projects 4 and 5, the scramble to pass on the hot potato of their costs had become a preoccupation among Northwest energy interests. What might be termed the establishment viewpoint was that the payments should be spread among a broader base of regional ratepayers, not just the customers of the eighty-eight participants. At a press conference in December 1982, following his resignation, Robert Ferguson candidly referred to a regional solution as a "bailout," a term that its proponents preferred to avoid.[6]

Public relations strategies sometimes prevailed in efforts to put the establishment viewpoint into effect. In the spring of 1983, Washington Governor John Spellman held a series of meetings with those worried about potential default. Aides discussed a "marketing strategy" to enlist supporters of schemes to meet the obligations on the terminated plants and a study designed to show that default would damage future infrastructure investment in the Northwest. Organizing "prayer" meetings to bring around potential opponents of regionalization was also on the agenda at these gatherings, although apparently this idea was soon dropped.[7]

Legally, the main arena was Chemical Bank's attempt to affirm the validity of the Participants' Agreements in Washington. King County Superior Court Judge H. Joseph Coleman was not exaggerating when he referred to

[5] Northwest Power Planning Council, *1983 Northwest Conservation and Electric Power Plan*, Volume 1 (Portland: The Council, 1983), 5–13. For an analysis of the region's surplus, see Marc M. Hellman, "The Economics of a Surplus in Electrical Generating Capability: The Pacific Northwest," *Public Utilities Fortnightly* 113, 1 (January 5, 1984): 45–47.

[6] *Clearing Up*, December 3, 1982, 9.

[7] U.P.I., Regional News, BC cycle, September 13, 1983.

the Chemical Bank proceedings as "an unusually contentious litigation."[8] Nothing came easily, as attorneys volleyed motions and counter-motions on normally mundane procedural matters like production of documents. Several utilities sought to bring Bonneville into the case, hoping to show that BPA had coerced them into signing the 1976 Participants' Agreements. "BPA is the main actor in this play, and you can quote me on that," insisted Tacoma attorney Al Malanca.[9] The U.S. Department of Justice attempted to get the case transferred to Federal District Court, but in December 1982 the federal court turned down the motion and the case stayed under state jurisdiction.[10]

Judge Coleman's rulings from the bench in the Chemical Bank case in October 1982 had upheld the authority of Washington utilities to enter into the 1976 Participants' Agreements. In December, he presented another set of summary judgments. Attorneys for the utilities had emphasized the defense that they had been the victims of misrepresentation at the time of the agreements. Supply System bond counsel Brendan O'Brien had conceded in a deposition that he and his colleagues had harbored doubts about the authority of sixteen of the participants and had not issued the customary legal opinion letters for these utilities. This opened the door for other Participants to contend that they had been lured into signing the pacts. While Coleman did not exclude all defenses based on misrepresentation, he held that O'Brien's concession did not invalidate the agreements. The judge also ruled that the utilities had not been coerced into signing the contracts and that the "hell or high water" clauses in the contracts did not make them "substantively unconscionable."[11]

Judge Coleman made it clear his rulings were only preliminary. "A trial is not only warranted but necessary," he announced in mid-December, setting a date of January 10, 1983.[12] At stake in the trial would be whether the Participants' Agreements had been reached by methods that were "procedurally unconscionable." For the Participants, a trial with Coleman presiding augured a verdict in favor of Chemical Bank and the bond

---

[8] *Clearing Up*, December 10, 1982, 10.

[9] Quoted in *Clearing Up*, October 22, 1982, 6.

[10] "WPPSS Suit Returned to Court in Wash. County," *The Bond Buyer*, December 2, 1983, 3.

[11] Howard Gleckman, "Wash. WPPSS Trial Narrowed to Issue of Misrepresentation," *The Bond Buyer*, December 16, 1982, 1. James S. Granelli, "The $7 Billion Default Threat," *National Law Journal*, May 30, 1983 offers a good summary of the case in Judge Coleman's court.

[12] Gleckman, "Wash. WPPSS Trial Narrowed...," 1.

investors, as his rulings had supported the plaintiff's claims. On January 25, 1983, the utilities were due to pay the Supply System their first installment of the funds which WPPSS, in turn, would pass on to Chemical Bank for disbursement to the bondholders on July 1, 1983. Not eager to start repaying the debt, the defendants wanted to postpone trial at least until after the first payment date. They also pressed for a speedy appeal of Coleman's summary judgments to the Washington State Supreme Court.

The participating utilities confronted the demand for their money with varying degrees of anger and resignation. Utilities with commissioners from the Irate Ratepayers movement adamantly rejected the notion that they owed the money. Others, especially utilities with ties to the Supply System, were less confrontational. The city of Richland, Washington, home to the headquarters of WPPSS, not surprisingly fell in the latter category. It proposed an escrow fund to hold Participants' payments pending a resolution of the case and, when Judge Coleman approved the arrangement, transferred $331,000 into the account. Most of the Participants, however, were less forthcoming. The tally at the end of the day on January 25 was two small utilities paying the Supply System, six making deposits to escrow, and four setting aside their own special funds. The rest were balking at making any gesture toward payment for the terminated projects. In the words of a representative of the Clark County PUD, "We're not doing anything." The Participants' choices were difficult; utility officials "are going through hell right now," said Jim Boldt, executive director of the Washington Public Utility Districts Association.[13]

While Chemical Bank pressed for prompt payment for the bondholders, utility interests grasped for legal strategies to invalidate the Participants' Agreements. A week before the utilities refused to pay the first installment, Snohomish PUD, the largest participant in the terminated plants, and Clallam PUD sued the Bonneville Power Administration in the U.S. Court of Claims, charging BPA with improperly pressuring them during the mid-1970s into participating in Projects 4 and 5. The suit also alleged that Bonneville had failed to oversee construction of the net-billed projects adequately. Eventually, the utility plaintiffs contended, cost escalation and construction delays on the net-billed plants saturated the bond market and forced termination of WNP-4 and 5. This element of the "seduction suit" (as the action came to be known) raised an additional specter for the Supply System. If the utilities could establish their assertion that BPA's

---

[13] *Clearing Up*, January 28, 1983, 7; U.P.I. Regional News, BC cycle, January 25, 1983; Howard Gleckman, "86 of 88 WPPSS Participants Ignore Debt Payment Deadline," *The Bond Buyer*, January 26, 1983, 1.

misdeeds regarding Projects 1, 2, and 3 had brought down the terminated plants, this might breach the barrier between the net-billed and the
participant-financed projects. Creditors on WNP-4 and 5 then might be
able to take assets from the net-billed plants to satisfy obligations arising
from the terminated projects. Maintaining the "Chinese Wall" between
the two sets of reactors became an intense concern for the leaders of
WPPSS. Although they were allied with the bondholders in their desire to
see the Participants pay for the terminated plants, on the "Chinese Wall"
issue Supply System and investor interests sharply diverged.[14]

The net-billed plants had problems of their own. Although Initiative
394 requiring voter approval for future WPPSS bond sales had been invalidated in June, the initiative's proponents were appealing the decision.
Public hostility to the Supply System had continued to grow. When the
Seattle Chamber of Commerce commissioned a survey of attitudes about
WPPSS, the polling firm's head reported, "Boy oh boy, were the impres-
sions negative."[15] In such a climate, and as long as the initiative had not
been firmly crushed, the Supply System understandably lacked confidence
that it could win a majority of voters for a new round of bonds. Nor were
underwriters eager to take on a new issue from the beleaguered borrower.

In early 1983, Project 3, over two-thirds complete, required as much as
$960 million to finish. By May 1983, its cash reserves would be exhausted.
Blunt-talking Carl Halvorson told the press in January that he was worried
about WNP-3 "because that takes a hell of a wad of dough." However, a
construction slowdown at the Satsop plant would be costly – "like piling
up a whole pile of $100 bills on the beach and then taking a D-9 Cat
[a bulldozer] and pushing them over."[16]

If the Supply System decided to put WNP-3 in mothballs, it could expect
to face the wrath and the attorneys of the four private utilities who owned
30 percent of the project. A year earlier, the IOUs had warned that they
would view a construction halt on Project 3 as a breach of contract (see
chapter 5). Now, the Executive Board temporized. On February 25, 1983,
WPPSS approved a four-day work week on the Project 3 site in order to
slow down the cash outflow, but kept the entire 3,900 person-workforce
on the job. The plan was intended to keep construction going with current

[14] On the politics of the suit against Bonneville, see *Clearing Up*, January 21, 1983, 7–8
   and U.P.I. Regional News, BC Cycle, January 19, 1983.
[15] Joel Connelly, "Public split on WPPSS debt, poll shows," *Seattle Post-Intelligencer*, March
   24, 1983.
[16] Bob Lane, "New WPPSS board chairman worries about Project 3, not 2," *Seattle Times*,
   January 17, 1983; Joel Connelly, "WPPSS is nearing 'brink' fast, its chairman warns,"
   Seattle Post-Intelligencer, January 21, 1983.

funds through August, but reports filtered out a week after the decision that the savings would not suffice to extend construction even that long.[17]

In late March, Peter Johnson, the Bonneville Power Administrator, stated that he hoped that WPPSS could re-enter the bond market by late May or June and borrow as much as $450 million to continue work on WNP-2 and 3. At the same time, he gave formal assurances to the Supply System that BPA itself would directly finance the rest of Project 2, already 94 percent complete, if the System proved unable to sell bonds.[18] Bonneville then would charge the net-billed project participants for their share of the costs. For WNP-3, however, Bonneville made no such promises. It seemed foolhardy for Bonneville to pledge nearly a billion dollars for a plant that would compound the region's surplus power problems.

Two other factors also hindered efforts to finance continued work on Plant 3. First, Wall Street perceived a risk that the Supply System would declare bankruptcy. This would cast all involved onto the murky waters of municipal bankruptcy law and would throw into doubt the Supply System's commitment to repaying bondholders for the net-billed plants. The Washington State legislature that spring was considering bills to bar WPPSS from declaring bankruptcy as a system, but the Supply System and many legislators wanted to preserve the possibility that declaring Projects 4 and 5 bankrupt would help to implement a regional compromise for paying the bonds on the terminated plants. By late May, legislation to prevent bankruptcy was doomed.[19]

A second obstacle to financing completion of WNP-3 was another legal case coming from Oregon. This one, labeled Springfield III, challenged the

[17] *Clearing Up*, March 4, 1983, 10.
[18] Johnson's pledge reported in "WPPSS Loses One, Wins One in Northwest Courts," *The Bond Buyer*, April 4, 1983. WNP-2, its past construction travails apparently purged from memories, had become the nuclear Northwest's pride and joy. Construction was moving rapidly, the plant was 94 percent complete, and fuel loading slated to begin that September. WPPSS anticipated commercial operation by February 1984. In the end, however, the first fuel was inserted near Christmas and full-power commercial operation began on December 12, 1984.
[19] The bankruptcy issue can be followed in the following articles: Howard Gleckman, "WPPSS Sends Bills to 4, 5 Utilities, Warns of Default," *The Bond Buyer*, April 8, 1983, 1; "Public Utilities Cool to Spellman Plea on WPPSS," *The Bond Buyer*, May 11, 1983, 1; Howard Gleckman, "WPPSS Market Hurt by Rating Cut and Possible Default," *The Bond Buyer*, May 17, 1983, 1; Howard Gleckman, "WPPSS Bills Hit Snags in Congress, Wash. Senate," *The Bond Buyer*, May 23, 1983, 1; "Judge Says Failure to Pay Won't Mean WPPSS 4,5 Default," *The Bond Buyer*, May 26, 1983, 1; Howard Gleckman, "BPA tells WPPSS Mothballing Plant 3 Is Only Option Left," *The Bond Buyer*, May 27, 1983, 1.

validity of the net-billed contracts for WNP-1, 2, and 3. In early April, underwriters formally notified the Supply System that without a ruling upholding the net billing agreements or some other assurance for investors no further bond financing would be possible.[20] When Federal District Court Judge James Redden ruled on Springfield III on April 27, he gave the Supply System a partial victory, holding that all utilities had authority to enter into the net billing contracts. Through the agreements, the utilities had purchased power from Bonneville, something legally permissible. Redden thus removed one of the obstacles to selling bonds to keep WNP-3 going. But at the same time he implicitly raised another problem. If the net-billed plants fell through, he stated, the federal government, through Bonneville, would "either have to 'eat' the loss ... or BPA will have to raise its power rates...."[21] A substantial BPA rate increase from a failed plant would surely encounter political resistance. At the same time, it would reduce electricity demand. If demand proved to be elastic, an increased rate per kilowatt-hour would actually lower Bonneville's revenues. This could precipitate a "death spiral" of ever-higher rates, shrinking demand, and declining revenues.

In May, Bonneville, facing its own $350 million revenue shortfall for the year, played its cards. In a letter to Carl Halvorson, Peter Johnson stated that unless the Supply System could devise a workable financing plan by May 27, it would have to slow or stop work on WNP-3. Noting the "laborious pace" of litigation, he stated that BPA would not promise to finance continued work when WPPSS's funds on hand ran out. The next day, Standard & Poor's suspended its bond ratings on all the net-billed projects. S&P wondered if the Supply System could "resist pressure to file under Chapter IX of the federal bankruptcy code." Mixing his metaphors, Halvorson lamented that WPPSS was barefoot "among the barracudas" and "in the snow at Valley Forge."[22]

At its May 27 meeting, the Supply System grudgingly accepted the inevitable and enacted a mothballing plan for WNP-3. The Executive Board voted a three-year construction suspension but delayed full

[20] *Clearing Up*, January 28, 1983, 5; *Clearing Up*, March 25, 1983, 1; *Clearing Up*, April 15, 1983, 5; Howard Gleckman, "WPPSS Sends Bills to 4, 5 Utilities, Warns of Default," *The Bond Buyer*, April 11, 1983, 1.

[21] City of Springfield v. Washington Public Power Supply System, et al. (Oregon, 1983), 564 F. Supp. 90, 95.

[22] "Wash. High Court Refuses to Lift Delay on WPPSS Contracts Trial," *The Bond Buyer*, May 13, 1983, 3; Howard Gleckman, "WPPSS Prepares for Default as S&P Suspends Ratings," *The Bond Buyer*, May 16, 1983, 1; *Clearing Up*, May 20, 1983, 7.

implementation for thirty days in hopes of positive financial developments. After the Board's vote, members of the anti-nuclear Light Brigade celebrated in the rear of the meeting room by noisily toasting their victory with champagne. The Board set out to look for "unconventional" lending sources – "pawn shops" in Cyrus Noë's sardonic phrase. Those shops were not open. By the end of July, almost half of the 3,000 workers at the Satsop project had been laid off.[23] Now WNP-2 was the only WPPSS project underway.

One of the few bright spots for the Supply System was finding a successor to Ferguson. In mid-May, Don Mazur, formerly Director of Projects, assumed the position of Managing Director. Mazur accepted a salary equal to Ferguson's, $125,000, along with his predecessor's joking assurance that the job was a "piece of cake." Ferguson had brought Mazur into the Supply System when he arrived in 1980, and the two had once worked together at Hanford's Fast Flux Test Facility. Mazur may have reflected the Supply System's turn away from provincialism (although one reporter claimed he looked more accustomed to "western-cut clothes and cowboy boots...than a Brooks Brothers suit"), but he shared with many old-timers in the utility community faith that "The power is needed."[24] Ferguson moved to a presumably less stressful position as Chairman of UNC Nuclear Industries, the contractor that operated the N-Reactor for the Department of Energy. His move to a facility that was literally adjacent to the Supply System's turbine-generator symbolized once again the intimate connections between the military and civilian nuclear operations and the blurred lines between public and private undertakings.

The Roof Falls In

On Wednesday, June 15, 1983, in the words of *Clearing Up* editor Cyrus Noë, "the roof fell in" on efforts to make the utility Participants pay for the terminated nuclear projects.[25] In a 7–2 decision, the Washington State Supreme Court overturned Judge Coleman's rulings. For the justices in the majority, the crucial question was whether Washington

[23] *Clearing Up*, May 27, 1983, 7 and June 3, 1983, 3; Martin Heerwald, "WPPSS to mothball fourth plant," U.P.I. Regional News, May 27, 1983.
[24] Wanda Briggs, "Don Mazur named new WPPSS chief," *Tri-City Herald*, May 13, 1983; Briggs, "New WPPSS chief: 'The power is needed'," May 14, 1983.
[25] *Clearing Up*, June 17, 1983, 2.

utilities had possessed a right to enter into the "take-or-pay" contracts they signed in 1976. Although Coleman had held that municipalities and public utility districts possessed authority, Justice Robert Brachtenbach, writing for the majority, rejected the contention.[26] Washington participants did have statutory authority to buy electricity for sale and distribution. Under the statute creating WPPSS, the utilities were also explicitly permitted to contract with operating agencies like the Supply System to buy or sell electricity. However, Brachtenbach maintained, the participants had contracted to buy something other than electricity, namely "project capability." They exchanged an unconditional guarantee to pay off the bonds WPPSS issued for a percentage of any power the projects *might* generate. "The unconditional obligation to pay for no electricity is hardly the purchase of electricity." In some states, such as South Carolina, municipalities had been given statutory authority to enter into obligations of this type, but Washington was not among them. "The participants simply are not authorized to guarantee another party's ownership of a generating facility in exchange for a possible share of any electricity generated," he asserted.[27] Significantly, although Judge Coleman in the trial court had ruled on several issues of authority and contract, the majority opinion stopped here and did not address any issues beyond whether the utilities had authority to enter into the Participants' Agreements. Because the agreements were *ultra vires*, beyond the utilities' powers, they did not impose a current duty to pay for the terminated projects. On top of his legal decision, Judge Brachtenbach added his view that the Agreements had been unwise policy. "[T]hey constructed," he commented, "an elaborate financial arrangement that required the participants to guarantee bond payments irrespective of whether the plant was ever completed; to surrender ownership interest and considerable control to WPPSS; and to assume the obligations of defaulting participants."[28] In a concurring opinion, Justice Fred Dore took Brachtenbach's criticism farther: "From reading the record, it is clear that the monumental crisis brought on by WPPSS was created by the simultaneous construction of five nuclear plants, as well as mismanagement. To save itself, WPPSS has asked us to approve a

---

[26] This Chemical Bank appeal directly involved only Washington State PUDs and municipal utilities, twenty-eight of the eighty-eight participants, but these included almost all of those with substantial shares of project capability.

[27] Chemical Bank v. Washington Public Power Supply System, 99 Wn. 2d 772 (1983), 782, 799.

[28] Ibid., 798.

plan to mortgage the futures of ratepayers by requiring huge increases in electricity rates, in exchange for nothing, in violation of our statutes and state constitution. This we cannot do."[29]

Two justices, Robert Utter and James Dolliver, dissented. Utter maintained that statutes granting powers to municipal utilities were to be interpreted broadly. The Participants' Agreements, he contended, were purchases of the "possibility of power." Though atypical, these contracts could have had advantages for the utilities. In the mid-1970s, indeed, predictions of rising energy prices and shortages had made acquisitions of this sort look prudent. At the time, the state legislature, rather than worrying that the utilities had exceeded their authority, had expressed no disapproval. But Utter also had harsh words for the "enormity of the mismanagement in this project and the calamitous impact of its failures on utilities and ratepayers."[30]

Responses to the Supreme Court's Chemical Bank decision were predictable. Utility boards now in the hands of angry ratepayers rejoiced that the court had spared them from the rate increases that would have required them to pay $7 billion in principal and interest to bondholders. Within two days of the verdict, Snohomish PUD, the largest participant, withdrew a 16.7 percent increase in consumer rates and also voted to request the return of funds it had put into escrow while the case was on appeal.[31] Other participants soon followed. But the public utilities could not feel entirely comfortable. Jim Boldt warned that Chemical Bank would come after anybody involved with WPPSS "like a pack of wolves."[32]

Wall Street and the Supply System's creditors were dismayed at the Chemical Bank decision, but some held out hope for a solution that would involve repaying the debt. One partner in a municipal bond firm termed the decision "astonishing" but said he did not consider it "the end of the line." The bond market responded with a drop of ten to fifteen points in the price of Project 4 and 5 bonds, which were already selling at deep discounts. Although the decision had no immediate impact on the legal status of the bonds for the net-billed plants, they declined by three points or more.[33]

[29] Ibid., 809–810.
[30] Ibid., 810–814. "Enormity" quote at 810.
[31] Howard Gleckman, "WPPSS Utility Seeks Return of Escrow Payment," *The Bond Buyer*, June 17, 1983, 1.
[32] Brian Mottaz, "Utility official says Chemical Bank will not rest," U.P.I., June 4, 1983.
[33] Howard Gleckman, "Wash. Supreme Court Rules Utilities Did Not Have Power to Sign WPPSS 4,5 Contracts," *The Bond Buyer*, June 16, 1983, 1.

Others were more alarmed about the regional effects. One brokerage executive predicted that the Supply System's looming default would spill over onto other Washington borrowers' credit evaluations: "The market will have to distinguish Washington from the other states." Governor John Spellman and other politicians understandably feared for the state's financial reputation and its ability to borrow. They intensified their search for a way to settle the WPPSS 4 and 5 debt. Spellman immediately renewed his calls for regionalization. This would require federal action to permit Bonneville to assume the debt on the terminated projects and collect the revenue from rate increases throughout the Northwest. Donald Mazur of the Supply System, endorsed this. "The only hope we have," he stated, "is federal help." But it was soon obvious that the Northwest congressional delegation would not endorse federal aid. Spellman himself conceded that Washington, D.C., was not going to provide funds. "There is not going to be a federal bailout. That is clear," he admitted.[34]

In later legal analysis, the Washington State Supreme Court decision in the Chemical Bank case has not fared well. Legal commentators have generally found it far too convenient that a Pacific Northwest court invalidated the Participants' Agreements and released the utilities from their obligations to pay. The wording of the agreements seemed explicit: "The Participants shall make the payments to be made to Supply System under this Agreement whether or not any of the Projects are completed, operable or operating. . . . " Utilities elsewhere had entered successfully into similar "take-or-pay" contracts. The language of legal scholarship on the Chemical Bank decision indicates the dominant academic reaction. The decision was "questionable," an "aberration," a case of "judicial meltdown," and its narrow approach should raise a "cry for reform."[35] Criticism focused on the court majority's strategy of construing utilities' powers narrowly. In general, "Dillon's rule," the prevalent doctrine about the powers of

---

[34] Ibid.; "WPPSS Bailout Plans Garner Little Support," *The Bond Buyer*, June 16, 1983, 1; Howard Gleckman, "Wash. Governor Told Congress Will Not Back WPPSS Bailout," *The Bond Buyer*, June 22, 1983, 3.

[35] "Note: Chemical Bank v. Washington Public Power Supply system: The Questionable Use of the Ultra Vires Doctrine to Invalidate Governmental Take-or-Pay Obligations," 69 *Cornell L. Rev.* 1094 (1984); Grant Degginger, "Comment, Chemical Bank v. Washington Public Power Supply System: An Aberration in Washington's Application of the Ultra Vires Doctrine," 8 *U. Puget Sound L. Rev* 59 (1984); Robert L. Tamietti, "Chemical Bank v. WPPSS: A Case of Judicial Meltdown," 17 *Natural Resources Lawyer* 373 (1984); Richard Shattuck, "A Cry for Reform in Construing Washington Municipal Corporation Statutes – Chemical Bank v. Washington Public Power Supply System, 99 Wn. 2d 772, 666 P.2d 329 (1983)," 59 *Washington L. Rev.* 653 (1984).

municipal corporations (a category that included the participants), has held since the 1870s that these bodies have only those powers expressly granted to them and that controversies over what they can do should be resolved in the negative. Commentators, however, have pointed out that the rule normally applies to the "governance" functions of municipalities, not to their "proprietary" activities, where broader powers are merited. Narrow construction is appropriate for the exercise of powers involving elements of sovereignty, they maintained, but not when the municipal body is attempting to advance the particular interests of its constituents, as in a public utility. For those activities, in Washington as in other states, municipal powers more closely resembled the broader authority that private corporations had. They deserved a more liberal interpretation.[36]

The commentators objected to Chemical Bank v. WPPSS on other grounds as well. For example, Judge Brachtenbach had claimed that the Participants' Agreements did not provide the utilities a true ownership interest in the projects. For this he used a concept of ownership that the critics called too restrictive. In their view, the agreements had given several elements of effective control to the utilities, and thus fell under statutes that allowed them to own electrical facilities. In the same vein, critics maintained, the court majority had looked at fragments of state statutes in isolation when it could and should have found a clear pattern of legislative intent to allow public utilities to enter into contracts like the Project 4 and 5 agreements. In the harshest critique, Robert L. Tamietti maintained that the Washington Supreme Court decision had been a "smokescreen" to protect local interests. The judges had presented "a picture of our judiciary at its worst," and had engaged in an "intellectually dishonest ploy" aimed at "sticking it" to bondholders instead of ratepayers.[37]

Some Northwesterners shared the commentators' sentiments. Even a decade later, Richard Quigley, who had been the Supply System's General Counsel, responded with emotion when asked how he had reacted to the Chemical Bank decision. "Nothing could be in our opinion further from right, irrespective of how that decision was written . . . that decision was absolutely wrong." Quigley had long admired Jack Cluck, "a lion among lawyers." Cluck was the public utility attorney whose Seattle firm,

---

[36] For comments on Dillon's Rule and the governance-proprietary distinction, see Hugh D. Spitzer, "Municipal Police Power in Washington State," 75 *Washington L. Rev.* 495 (2000), 495–498.

[37] Tamietti, "Judicial Meltdown," 381, 395.

Houghton Cluck Coughlin and Riley, represented WPPSS and many other Northwest public utilities for many years. Cluck passed away soon after the decision. "When the Supreme Court of this state came down with the decision that WPPSS [sic] didn't have the authority... I suspect that's what killed Jack Cluck."[38]

Yet however debatable the legal reasoning of the majority opinion in the Chemical Bank case, as editor Cyrus Noë observed at the time, the decision "validates the legal actions of various utilities in answering Chemical Bank's request for declaratory judgment with a challenge to authority." Whether the decision "makes good law or bad law, it demonstrated (if it really needed demonstrating) that legal resistance by the utilities was well grounded and serious...."[39] Noë struck a sensible balance. However sensitive the court's majority may have been to regional ratepayer interests (and critics presented no evidence to back insinuations of bias or dishonesty), the participating utilities helped their ratepayers by gaining a ruling that promised to save them literally billions of dollars. Not to seek relief from enormous outlays for the two abandoned projects would have been irresponsible to their customers.

By claiming they had lacked authority to enter into the agreements, the utilities had managed to stave off financial crisis. However, they risked undercutting their own capacity for effective action in the longer run. As critics of the Chemical Bank decision pointed out, joint operating agencies (JOAs) like WPPSS had been established around the country to allow public entities to band together to accomplish what they could not achieve as individual units. The take-or-pay clauses in the agreements they signed were designed to reduce interest rates for agency debt by pledging the income sources of the utilities (their ratepayer revenues) to repay the JOA bonds. If court decisions confined public utilities to the powers explicitly granted them by state statutory law, would they be able to respond effectively to a shifting environment?

Compounding both the irony and the insecurity of the utilities' position, the Chemical Bank decision, delivered during the first term of the Reagan administration, coincided with a drive for privatization of government functions. If the court had required Participants to pay for the terminated plants, ratepayer anger would have further weakened the public utilities. But escaping the debt had its own political price. The court had

---

[38] Interview by author, Kennewick, Washington, October 15, 1992.
[39] *Clearing Up*, June 17, 1983, 6.

construed the authority of the Participants very narrowly. If public utilities, the products of movements for governmental activism, found their powers curtailed, might this not also feed antigovernment sentiment? In 1984, the President's Private Sector Survey on Cost Control, known as the Grace Commission Report, proposed that the federal government sell its power marketing agencies, including BPA, and the federal-owned dams on the Columbia.[40] Proposals to sell Bonneville gained some support, although most of the Northwestern congressional delegation fought against them.[41] The following year, David Stockman, head of the Office of Management and Budget, called for refinancing at higher rates the debt BPA owed the federal government. Stockman maintained this would end an unfair subsidy to Northwest ratepayers. Again regional politicians managed to block this, but again public power interests in the Pacific Northwest had been challenged.

The fate of the state Supreme Court's Chemical Bank decision indicates that the majority viewpoint fell within the boundaries of mainstream jurisprudence. That verdict had applied to Washington State municipal utilities and Public Utility Districts. Returning to Judge Coleman's trial court in August 1983, the Participants not directly affected by the state Supreme Court's ruling contended that they too were no longer bound to the Participants' Agreements. Coleman agreed. With utilities holding 70 percent of the project shares released, the contracts could not legally or practically be enforced against the other participants. The judge therefore ruled that none of the eighty-eight utility participants was legally obligated to pay for WNP-4 and 5.[42]

Chemical Bank appealed Coleman's August ruling back to the Washington Supreme Court and also asked for reconsideration of the court's June 15 decision, but the case moved slowly this time. In a 6–3 verdict on November 6, 1984, nearly a year and a half after their initial decision, the Washington State Supreme Court upheld its previous *ultra vires* ruling.[43] It also affirmed Judge Coleman's order releasing the utilities not covered in the initial case. No doubt frustrating for Chemical Bank, its attorneys, and the bondholders was the fact that all of the justices who had voted

---

[40] *War on Waste: President's Private Sector Survey on Cost Control* (New York: Macmillan, 1984), passim and 437.

[41] The chief exception was Jim Weaver, who crusaded in favor of a plan to have Bonneville sold to the Northwest states. See Tollefson, *BPA and the Struggle*, 410–415.

[42] "All Participants in WPPSS 4, 5 Freed from Debt," *The Bond Buyer*, August 10, 1983, 1.

[43] 102 Wash. 2d 874; 691 P.2d 524.

to invalidate the agreements in 1983 and remained on the court voted as they had previously. The third dissenter in 1984 was a new member of the bench.

Chemical asked for a reconsideration of the Washington appeal decision, but the court turned this down six weeks later. The bond trustee, joined by the Supply System itself, then asked the United States Supreme Court for certiorari in early 1985, in effect requesting reversal of the state decision, but in April 1985 the Court denied the petition without comment, terminating judicial consideration of the case. Six months earlier, the justices had also upheld an Idaho Supreme Court decision that Idaho municipals had lacked authority to enter into the Participants' Agreements. A generally conservative U.S. Supreme Court, with seven of its members appointed by Republican presidents, had twice in half a year rejected pleas to overturn decisions that bondholders and their allies felt violated basic property rights and the sanctity of contracts. Admittedly, the Supreme Court is seldom eager to impose its own interpretations of state statute law, but its failure to take up either the Idaho or the Washington case further indicates that the state courts were not merely catering to ratepayers when they ruled that the Participants lacked authority. "Chapter one is closed. There's no doubt about it," conceded William Berls, the Chemical vice-president leading his bank's involvement in the Supply System case, after the U.S. Supreme Court had rejected the appeal.[44] The story, however, had by no means ended. The Supply System's saga was becoming a legal epic of almost-Dickensian complexity and duration – and with higher stakes.

The Chemical Bank decision destroyed the possibility of further borrowing for WNP-3. Don Mazur traveled to New York for discussions with underwriters late in June, and the Supply System presented a budget for continued construction on the project to BPA, but the talks produced nothing and Bonneville quickly vetoed the budget plan. On July 8, 1983, the Executive Board implemented an "immediate extended construction delay," expected to last for three years. Some recognized that delay might turn into "slow death." Governor Spellman had publicly worried that "The history of this country is that any plant put on hold is never started again."[45]

[44] *Clearing Up*, May 3, 1985, 7–8.
[45] Howard Gleckman, "WPPSS Halts Building of No. 3; To Sell 4,5 Assets," *The Bond Buyer*, July 11, 1983, 1; Brian Mottaz, "Governor believes federal government can help WPPSS," U.P.I. BC Cycle, June 22, 1983.

Spellman returned to his strategy of two years earlier, a blue-ribbon committee. Oregon's Governor Victor Atiyeh joined in the venture. The new group, charged with recommending solutions to the financial crisis, looked familiar. Spellman named Charles Luce, who had served briefly on the reconstructed WPPSS Executive Board, Edward Carlson, a member of Spellman's 1981 panel on cost overruns, and Herbert Schwab, retired Chief Judge of the Oregon Court of Appeals. Other Northwest dignitaries had reportedly declined to serve on this Advisory Panel.[46]

Default

It was from this condition – two plants terminated and two in mothballs in the summer of 1983 – that the Washington Public Power Supply System fell into the largest municipal bond default to that point in American history. Anticipated for months, default itself came as an anticlimax following the series of blows that the Supply System had absorbed in the previous three years: the 1980 Washington State Senate Inquiry into cost overruns; the construction halt on Projects 4 and 5; passage of I-394; termination of the two projects; the construction stoppage on WNP-1; court decisions in Oregon, Idaho, and Washington; the suspension of bond ratings; exclusion from the bond market; and the mothballing of WNP-3.

Even before the June 15 state Supreme Court decision, default loomed as almost inevitable. The Supply System had managed to make monthly payments of about $15.6 million to Chemical Bank through April, but at a May meeting the Executive Board decided it had to devote its shrinking cash on hand to paying administrative and legal costs. With the Supply System embroiled in sixty-six active cases, the logic of the Board's choice was easy to understand. To rub salt in its wounds, the bond covenant for the terminated projects seemed to obligate the Supply System to pay legal expenses for Chemical Bank's efforts to gain payment from the utility participants for the bondholders. WPPSS was considering boosting its in-house legal staff. So many attorneys had been hired to pursue actions against it that few suitable law firms were available to represent the Supply System. By late June, eighty-three law firms were involved in bondholders' suits. "It's hard to find firms that aren't already representing someone," noted a member of the WPPSS legal staff.[47]

---

[46] Gleckman, "WPPSS Halts..."

[47] Howard Gleckman, "WPPSS Suspends Work on Plant 3 for Three Years," *The Bond Buyer*, May 31, 1983, 1; "Firms in Bondholders' Actions," *Legal Times*, June 27, 1983,

Judge Coleman intervened to prevent a default in May, issuing an injunction that prevented Chemical Bank from declaring WPPSS in default while the Supreme Court was considering the case. To nobody's surprise, the Supply System again missed its payment to the bond trustee at the end of June. This did not prevent bondholders for WNP-4 and 5 from receiving their July 1 semi-annual interest payments. Chemical Bank held enough in its interest account to pay the $94 million due. However, following the state Supreme Court decision, this looked like the last coupon bondholders would be able to clip. Chemical's next step was to go before Judge Coleman with a request to lift his injunction and permit it to initiate default proceedings. On July 22, Coleman removed the stay and allowed the default process to get underway. Later that day, Alexander Squire, the Supply System's Deputy Manager, sent notice to the bank that it was unable to make payment. When Chemical received the letter on the following Monday, July 25, it demanded transfer of any remaining WPPSS funds. The Supply System immediately complied and handed over the $25.7 million left in the WNP-4 and 5 account. Chemical's Vice-President Berls told the press that the bank would soon take the final step in the default process, an "acceleration" notice demanding immediate payment of the entire debt. He also indicated that the bank would try to breach the "Chinese Wall" and tap into Supply System assets and revenues from the net-billed projects.[48]

The process of default was almost automatic. In fact, Supply System staff handled the mechanics of default without even notifying the Board of Directors or the Executive Board. However, after expressing unhappiness at their exclusion, Board members granted that managers had done what they had to. One of the reasons why WPPSS hurried through the default was another looming financial blow. In 1981, the Supply System had sued a uranium supplier, Western Nuclear, Inc., for antitrust violations; Western Nuclear counterclaimed that WPPSS was attempting to break a fuel supply contract. A federal judge had ruled in early 1982 that the supply contract was valid and initiated a long period of calculation of the damages WPPSS would be ordered to pay the vendor. By the summer of 1983, WPPSS officials feared a large judgment was imminent and that Western Nuclear might be allowed to seize assets, including any money

---

12ff; "WPPSS Looks for Ways to Control Its Rapidly Growing Legal Expenses," *The Bond Buyer*, June 3, 1983.

[48] Howard Gleckman, "Wash. State Courts Lift Final Barriers to WPPSS Default," *The Bond Buyer*, July 25, 1983, 1.

sitting in the WNP-4 and 5 account. Thus, the rush to pay Chemical what it had on hand was in part an attempt to avoid paying Western Nuclear. Indeed, within days of the default, Western Nuclear won a $53.6 million judgment against WPPSS.[49]

Chemical Bank intensified its legal struggle for the bondholders. On August 3, the bank filed suit for fraud, negligence, and breach of contract against the Supply System and named an extraordinary roster of 500 organizations and individuals as co-defendants, along with another hundred "John Does." Virtually everyone associated with the collapse of WNP-4 and 5 faced the bond trustee's challenge. Significantly, however, the bank decided not to name the bond underwriters, rating agencies, and bond counsel, "their colleagues and customers on Wall Street," as an anonymous source noted.[50] Otherwise, the bank's targets were similar to those of the bondholders who had been seeking to recover their money. It followed this up later in the month by issuing the "acceleration" notice. The federal Judicial Panel on Multidistrict Litigation issued an order on August 5, 1983, centralizing all the bondholder actions into one suit and adding Chemical to the list of plaintiffs. There was some controversy about the venue of pretrial proceedings. Some wanted to put the case in New York, but most parties preferred the federal Court for the Western District of Washington. Richard Bilby, an Arizona jurist, was assigned the case. (Federal judges in Washington State probably were also utility ratepayers, raising issues about their impartiality.) Like Chemical's decision to exclude bond brokers and counsel from its defendants' list, the question of venue was to become troublesome. The case became known as Multidistrict Litigation 551, or MDL 551 to those who would remain enmeshed in it for most of the remainder of the decade.[51]

---

[49] WPPSS appealed the award. In April 1984, the parties reached an out-of-court settlement for $25 million. Bonneville Power Administration supplied the funds for the payment. The case can be followed in: Frank Pitman, "WPPSS Claims Cartel Antitrust Violations in Suit to Void Western Nuclear Contract," *Nuclear Fuel*, December 7, 1981, 4; Harriet King, "Federal Judge Says WPPSS Must Pay Damages for Refusing Uranium from Western Nuclear," *Nuclear Fuel*, March 15, 1982, 7; *Clearing Up*, July 29, 1983, 7–8; "WPPSS and BPA Settle Dispute Over Nuclear Fuel," *The Bond Buyer*, April 13, 1984, 3; "BPA Supplies $25-million Cash for WPPSS-Western Settlement," *Nuclear Fuel*, April 23, 1984, 5.

[50] *Clearing Up*, August 5, 1983, 9–10.

[51] "Senate Debates WPPSS Aid Bill; Bank Sues on 4, 5," *The Bond Buyer*, August 4, 1983, 1; In Re Washington Public Power Supply System Litigation, 568 F. Supp.1250 (1983), Judicial Panel on Multidistrict Litigation); *Clearing Up*, August 12, 1983, 10.

Consequences of Default

When debtors fail to pay their creditors, the latter are almost certain to be unhappy. When the debtor is a governmental body far removed from the lenders, and when the money is lent upon strong assurances of repayment, creditors are likely to be still angrier. In the 1840s, during a period of American state defaults on transportation project bonds, most notably the collapse of the ill-conceived Pennsylvania Main Line Canal system, the English author Sidney Smith fulminated on behalf of British bondholders, "I never meet a Pennsylvanian at a London dinner without feeling a disposition to seize and divide him. How such a man can set himself down at an English table without feeling that he owes two or three pounds to every man in the company, I am at a loss to concede; he has no more right to eat with honest men than a leper has to eat with clean men...."[52] Would creditors now ostracize the reprobate ratepayers of the Pacific Northwest?

The Washington Public Power Supply System, across the continent from the nation's center of finance, serving a population paying less than half of the national average rate for electricity, and possessing seemingly valid agreements with eighty-eight utilities to pay even for terminated projects, had not merely skipped an interest payment. The Supply System had, however unwillingly, carried out the largest American municipal default ever. The rest of the country would not worry much about the travails of Northwest ratepayers. As one Eastern attorney put it, "While some old lady in the East is sitting in her apartment in a sweater freezing, those bastards out there are lit up like Christmas trees."[53]

Nor could nuclear ventures expect much public sympathy. By the early 1980s, public discourse on nuclear power focused on the industry's failures. Protest demonstrations like the ones in the mid-1970s at Seabrook had waned, but the mainstream media were reporting regularly on nuclear project fiascos. The Tennessee Valley Authority, with nuclear construction aspirations that had outstripped even WPPSS, canceled plans for eight plants between 1982 and 1984. The LILCO (Long Island Lighting Company) plant at Shoreham, New York, was virtually complete by 1983, but

---

[52] Quoted in Louis Hartz, *Economic Policy and Democratic Thought: Pennsylvania, 1776–1860* (first published 1948; Chicago: Quadrangle Books, 1968), 19.

[53] Gail Diane Cox, "Hunt for a Culprit Continues; WPPSS: Round Two," *National Law Journal*, April 13, 1987, 1.

challenges to the project's evacuation plans prevented it from ever opening and eventually left a $6 billion slightly radioactive hulk to be disman-
tled in the 1990s. In 1984, Public Service Company of New Hampshire
decided to abandon its second nuclear project at the Seabrook site. The
same year, Public Service Company of Indiana gave up on two partly-
completed reactor projects at Marble Hill. Cincinnati Gas and Electric
halted its Zimmer plant when it was 97 percent complete, hoping to convert it into a fossil-fuel facility. Consumers Power of Michigan terminated
a plant at Midland that had soared $3.25 billion over budget.[54]

The title of a February 1984 study by Congress's Office of Technology
Assessment (OTA), *Nuclear Power in an Age of Uncertainty*, sums up the
peaceful atom's problems around the time of the Supply System's default.
Asked what Congress could do to "revitalize the nuclear option," OTA
staff and consultants returned a wary report. It found many obstacles
to the recovery of nuclear power and stated that "significant changes in
the technology, management and level of public assistance" were necessary, but probably not sufficient, conditions for its revival. Financial and
economic prospects were bleak. Doubt about the nuclear option came
from all sides – macroeconomic trends, ratepayer decision-making about
energy usage, rapidly rising capital costs and long construction delays,
and a difficult financial and regulatory environment.[55]

Financial uncertainty beset a broad swath of American business in the
late 1970s and early 1980s. To some eminent scholars, this represented a
remarkable change. Theorists from Adolph Berle and Gardiner Means to
John Kenneth Galbraith had posited that the modern American corporation was insulated from external financing needs, largely able to fund its
growth from retained earnings, and (in Galbraith's view) almost assured
a profit each year through its ability to manage the demand for its products.[56] But the experience of the post-oil shock decade had called these
beliefs into question. To many observers, American corporations seemed

---

[54] Valuable studies include Joan Aron, *Licensed to Kill? The Nuclear Regulatory Commis-
sion and the Shoreham Power Plant* (Pittsburgh: University of Pittsburgh Press, 1997)
and Henry F. Bedford, *Seabrook Station: Citizen Politics and Nuclear Power* (Amherst:
University of Massachusetts Press, 1990). A useful overview is *Nuclear Power in an
Age of Uncertainty* (Washington, DC: U.S. Congress, Office of Technology Assessment,
OTA-E-216, February 1984).

[55] *Nuclear Power in an Age of Uncertainty*, 4, ix and chapter 3, passim.

[56] Adolph A. Berle, Jr. and Gardiner C. Means, *The Modern Corporation and Private Property* (New York: Macmillan, 1933); John Kenneth Galbraith, *American Capitalism: The
Concept of Countervailing Power* (Boston: Houghton Mifflin, 1952) and *The New Industrial State* (Boston: Houghton Mifflin, 1967).

increasingly "hollowed out," lacking productive capacities. Growing international competition in many industries meant that corporations could not count on a steady stream of retained earnings for reinvestment. A profit squeeze that lasted from the late 1960s through the early 1980s necessitated a shift to debt financing. Conservatives and some liberals feared that government borrowing to cover federal deficits would "crowd out" productive private investment. Corporate debt burdens had doubled since 1969, and the average Standard & Poor's corporate bond rating declined steadily from the late seventies onward.[57]

Meanwhile, financial markets were exploding. New forms of financial investments entered the marketplace. Financial institutions and nonfinancial firms both adopted investment strategies far more adventurous than those they had pursued in earlier decades. New computer and telecommunications technologies enabled accelerated trading. Sober observers worried about the fragility of the system. One major bank's report on "Credit and Capital Markets 1983," stated that "major credit and capital markets are characterized by more stresses and strains than is typical at this stage of a business cycle." Even Paul Volcker, chairman of the Federal Reserve, fretted in 1985, "We spend our days issuing debt and retiring equity... and then we spend our evenings raising others' eyebrows with gossip about signs of stress in the financial system."[58]

The fiscal health of state and local governmental units was also suspect in the early 1980s. "Prolonged periods of high interest rates and stagnant economic performance that characterized the past decade have continued into the 1980s and have contributed to the fragile financial structure of our cities," commented one scholar.[59] New York City's financial crisis of the mid-1970s may have been *sui generis* but other municipalities shared problems. Suburbanization and the decline of older manufacturing

[57] See Mark J. Warshawsky, "Is There a Corporate Debt Crisis? Another Look," chapter 6 in R. Glenn Hubbard, ed., *Financial Markets and Financial Crises*, National Bureau of Economic Research Project Report (Chicago: University of Chicago Press, 1991), 221, 224–5.

[58] Economics Department, Bankers Trust Company, "Credit and Capital Markets 1983," January 28, 1983, 1. Volcker quoted in Anthony Bianco, "Get Rich Today Come What May," *Business Week*, September 16, 1985, 90. The account of the corporate economy in the 1980s in Bennett Harrison and Barry Bluestone, *The Great U-Turn: Corporate Restructuring and the Polarizing of America* (paperback ed.; New York: Basic Books, 1990) influences my treatment here. Also on the anxieties of the "casino society" in the mid-1980s is "Playing with Fire," *Business Week*, September 16, 1985, 78–90.

[59] James H. Carr, "Introduction," in Carr, ed., *Crisis and Constraint in Municipal Finance* (New Brunswick, NJ: Center for Urban Policy Research, 1984), x.

industries had compounded the ills of core cities, but stagflation and recession had also taken a toll on other municipalities and special districts like the Supply System. Commercial banks and property and casualty insurers that had lowered their corporate income tax bills by investing in tax-exempt securities saw their profits sink in the recession of the early 1980s and lost much of their incentive to shelter profits in the municipal market. The Tax Equity and Fiscal Responsibility Act of 1982 imposed several limitations on tax-exempt securities issues in ways likely to lower demand and raise interest rates for cities and other borrowers. High interest rates drastically increased the cost of public finance. Historically, a quarter of state and local debt service outlays had gone to interest payments and three-quarters to principal reduction. In the early 1980s, the proportions were reversed.[60]

The jittery condition of municipal and nuclear power finance in the early 1980s contributed fears that the WPPSS default would have severe repercussions. The day after default, in the fevered words of a *Washington Post* reporter, "the Pacific Northwest awoke...to find itself transformed into a huge financial disaster area, indelibly tainted by the abandonment of nuclear plants...." The press secretary of Governor John Spellman complained that the region was experiencing "guilt by association" that would raise interest rates for all Northwest municipal borrowers, "even schools and school districts." The ripples from the default might also spread outside the area. "Hell or high water" clauses in utility bond agreements were used throughout the country. That courts in all three of the Pacific Northwest states had found cracks in their seemingly impregnable barriers against default portended problems for similar contracts around the country.[61]

As events unfolded, however, anticipated trouble outstripped actual financial damage. Admittedly, when the state of Washington went to market to sell $150 million in general obligation bonds in early August 1983 – during the final act of the Supply System default drama – Wall Street estimated that the state was paying seventy-five to one hundred basis points higher interest than comparable bonds were yielding. The WPPSS effect probably accounted for at least three-quarters of the added interest burden for the state. A week later, when Washington borrowed $200 million in revenue anticipation notes (short-term securities that states use to raise

[60] Randy Hamilton, "The World Turned Upside Down: The Contemporary Revolution in State and Local Government Capital Financing," in Carr, ed., *Crisis and Constraint*, 200.
[61] Jay Mathews, "Default of WPPSS Taints Development in Pacific Northwest," *Washington Post*, July 26, 1983.

funds before tax revenues are due) the state paid an extra twenty-five to forty basis points – a harsh penalty on these routine, very safe notes.[62]

Nevertheless, negative impacts were limited and transitory. When a Michigan public utility sold revenue bonds in September, it found investors willing to buy at a lower interest rate than expected, and there was no sign of a WPPSS-related penalty. Even Pacific Northwest bonds seemed to shed their stigma quickly. In October, Washington sold another round of general obligation bonds at an interest rate half a percent less than the August issue and lower than some comparable bonds other issuers sold the previous week. On the other hand, individual utilities participants in the terminated WNP-4 and 5 or other Northwest nuclear ventures did appear vulnerable. Moody's that fall lowered the ratings of several regional utilities with investments in the Portland General Electric's Trojan plant. Snohomish PUD had to pay a premium when it sold bonds in November; its manager complained that bond rating agencies had cast a "regional wet blanket" over the Northwest's public utilities.[63]

Several econometric studies of the Supply System crisis and default are in agreement that neither the general municipal bond market nor the common stocks of electric utility companies suffered greatly. (Utilities with a high exposure to nuclear power did face lower returns and greater variability in the months following default than they had in 1982 before default loomed large. The weakening of utilities with the heaviest involvement in nuclear energy was probably a rational market response to the fundamental situation of these firms rather than a direct result of the Supply System default.) One of these analyses had found that the New York City near-default of the mid-1970s harmed a broad sector of municipal borrowers, but it showed no comparable impact from the WPPSS woes.[64]

[62] William J. Ryan, "Wash. GO Issue Slightly Better Than Half Sold at Yields out to 10.75% in '08," *The Bond Buyer*, August 9, 1983, 1; George Yacik, "Wash. Notes Penalized by 25 to 40 Basis Points Because of WPPSS," *The Bond Buyer*, August 16, 1983, 1.

[63] *Clearing Up*, November 4, 1983, 4; *Clearing Up*, November 11, 1983, 9; *Clearing Up*, November 23, 1983, 1.

[64] John W. Peavy III and George H. Hempel, "The Effect of the WPPSS Crisis on the Tax-Exempt Bond Market," *Journal of Financial Research*, 10, 3 (Fall 1987): 239–247; P.R. Chandy and Imre Karafiath, "The Effect of the WPPSS Crisis on Utility Common Stock Returns," *Journal of Business Finance and Accounting*, 16, 4 (Autumn 1989): 531–542; Richard L. Smith and James R. Booth, "The Risk Structure of Interest Rates and Interdependent Borrowing Costs: the Impact of Major Defaults," *Journal of Financial Research*, 8, 2 (Summer 1985): 83–94. One early nonquantified study did suggest that Northwestern municipalities and other nuclear utilities were paying a substantial price in their borrowings. See L.R. Jones, "The WPPSS Default: Trouble in the Municipal Bond Market," *Public Budgeting and Finance*, 4 (Winter 1984): 60–77.

The Wrath of the Rentiers

There are two radically divergent images of bondholders in modern capitalist societies. On the one hand, critics often have depicted them as impediments to economic progress, coupon-clippers in quest of high interest rates who pinch the supply of capital and hinder entrepreneurship. The nineteenth-century usage of "capitalist" to mean parasitic money-lender, contrasted with the productive "manufacturer," reflects this attitude.[65] John Maynard Keynes's suggestion of the "euthanasia of the rentier" encapsulates another negative view. Keynes contended that the rentier role would gradually fade away. His contemplation of a "somewhat comprehensive socialization of investment" reflected his belief that the workings of the private capital market could not ensure a full employment level of production.[66] The other image of the bondholder is that of the diligent saver who husbands society's seed corn and places it in the hands of those who will use it for growth. At the same time, this bondholder provides for her personal and familial future by assuring future income for old age, illness, or the education of children. Not constricting rentiers but conscientious "widows and orphans" are at the center of this picture.[67] The struggle over the WPPSS default was both a legal conflict and a battle over which image better portrayed the Supply System's bondholders.

Cyrus Noë periodically attempted to ascertain just who owned bonds in the terminated plants, but the task was not simple. Names supplied for legal discovery in the King County case in late 1982 revealed holders of only about $90 million, approximately 4 percent of the total, and held "no surprises." Shortly after the Chemical Bank decision, Noë compiled a list of institutional owners, which also was incomplete and without startling news. Even though commercial banks and property and casualty insurance companies had been turning away from tax-exempts in a period of low profits, and despite the likelihood that they would have better

---

[65] These remarks oversimplify a complex history of the term "capitalist." For a brief, incisive essay, see Raymond Williams, *Keywords: A Vocabulary of Culture and Society* (New York: Oxford University Press, 1976), 42–44. A broad intellectual history is Jerry Z. Muller, *The Mind and the Market: Capitalism in Modern European Thought* (New York: Knopf, 2002).

[66] John Maynard Keynes, *The General Theory of Employment Interest and Money* (London: Macmillan, 1936), 374–381.

[67] *Seattle Times* reporter Bob Lane sketched a personification of this image as default loomed, a seventy-year-old woman collecting her semi-annual WPPSS bond interest payment and heading off "to get ready for the grandkids." "WPPSS bonds and 'the little old lady in tennis shoes'," *Seattle Times/Seattle Post-Intelligencer*, July 3, 1983.

knowledge than individual investors of the risks of these bonds, these institutions owned large amounts of WPPSS debt. American Express held $90 million in WNP-4 and 5 bonds and $200 million in all the projects. Major insurers – Aetna Life and Casualty, CIGNA, and Fireman's Fund – all held $40 million or more of the terminated projects. State Farm, which showed no Project 4 or 5 bonds, owned over $250 million of net-billed Supply System issues. A small municipal bond brokerage in California faced a move by regulators to shut it down for violating net capital and customer reserve requirements. "We have just too many WPPSS bonds at the wrong time," confessed the company's president.[68]

Estimating the number of individual bondholders was complicated by the fact that brokerages had packaged many of the bonds into unit investment trusts (UITs). Chemical Bank stated that approximately 10,000 people owned bonds directly but that another 65,000 owned shares in UITs containing WNP-4 and 5 securities. Ultimately, about 25 percent of these bonds were held in UITs. One fund alone, the Municipal Investment Trust Fund sponsored by Merrill Lynch and three other firms, contained $232.5 million worth, 10.3 percent of all the WNP-4 and 5 bonds issued. At the end of 1982, John Nuveen, a municipal bond specialist broker-age, had another $131 million in its Nuveen Tax-Exempt Bond Fund series, 5.8 percent of the total, but it accelerated a policy of selling off those bonds in early 1983. (It reported disposing of all its WNP-4 and 5 bonds by August and its holdings in the net-billed projects by November, 1983.) Not surprisingly, following the default, investors in these funds sued Merrill Lynch and Nuveen, asserting that the firms had failed to disclose negative information about the projects.[69]

Some institutional buyers had been wary of WNP-4 and 5 bonds for years. Several insurance companies expressed doubts to Donald Patterson, the Blyth Eastman executive who served as financial advisor to WPPSS, as early as 1977. Property and casualty insurers held 32 percent of WNP-4 and 5 bonds at the end of 1978 but only 23 percent two years later. In many cases, however, high interest rates overcame any investor apprehensions. Putting the bonds into a unit trust had the appealing effect of pumping up

---

[68] *Clearing Up*, November 12, 1982, 2; *Clearing Up*, June 24, 1983, 8; David Zigas, "SEC Seeks to Close Gibralco after WPPSS Losses," *The Bond Buyer*, June 22, 1983, 1.

[69] SEC, *Staff Report*, 219; David Zigas, "Unit Trust Investors Bought 16% of All WPPSS 4,5 Bonds," *The Bond Buyer*, November 1, 1983, 1; Zigas, "Merrill Lynch Sued Over WPPSS Bonds Held in Unit Trust," *The Bond Buyer*, September 1, 1983, 1; Zigas, "Nuveen Sued for Fraud In Selling Unit Trusts With WPPSS Bonds," *The Bond Buyer*, July 17, 1984, 1.

the UIT's overall yields and making the trust's managers look successful. Moreover, Supply System bonds were such a dominant presence on the municipal utility bond market that adding them to a UIT was, if not unavoidable, on the path of least resistance.[70] Finally, as some brokerages became more worried, they seem to have concluded that they had to keep buying WNP-4 and 5 bonds to protect their earlier investments and stave off a crisis.[71]

A June 1983 *Clearing Up* tabulation of institutional holdings amounted to $372 million, or about 16.5 percent of the total amount borrowed for the terminated projects. If the list is nearly complete (admittedly a large assumption), then individuals held approximately five-sixths of the bonds, either directly or in UITs (or bond mutual funds, although these investment vehicles remained a minor force in the municipal market). With about 75,000 investors, this would mean that the average holder owned about $25,000 in the tainted bonds. The 1988 staff report of the Securities and Exchange Commission on WPPSS bond sales indicated that a quarter of the bonds were in UITs in May 1981, when construction was suspended. Since the trusts bought and held their bonds in a fixed portfolio, and since there were no further issues of WNP-4 and 5 bonds after May 1981, we can assume that this was approximately the proportion the trusts owned in 1983. The 65,000 investors who owned the troubled bonds through UITs thus probably held about $560 million, or an average of less than $9,000. The 10,000 who bought the bonds directly possessed roughly 55 percent of the total, with face value of $1.24 billion. By these calculations, their average investment would be about $124,000. An investor with $124,000 in bonds paying 12.44 percent (the net interest cost of the highest yielding issue) would receive about $15,400 in interest annually – if the Supply System had not defaulted. Although tax free and at 1983 price levels, this income was nowhere near enough to judge the average WPPSS bondholder to be a plutocratic rentier. It was enough, however, to indicate that the bondholders who had bought the securities outright rather than via a UIT were likely to be well-off. (For comparison, the median family income in the United States in 1983 was $24,580.)

---

[70] Gleckman, "WPPSS: From Dream to Default," 179–180; SEC, *Staff Report*, 227. Among the twenty-five joint action agencies issuing power bonds between 1975 and 1982, the Supply System was responsible for 42 percent of the total, more than five times as much as the next-largest issuer. "Public Power Bond Issues up 35%," *Public Power*, 41, 2 (March-April 1983): 63.

[71] Gleckman, "WPPSS: From Dream to Default," 179–183.

Clearly these WPPSS bondholders were not primarily frugal widows and struggling orphans.[72]

Another hint about the status of the bondholders came in a 1985 survey that their organization, the National 4/5 Bondholders' Committee, conducted. For obvious reasons, the investors reported the results in a manner designed to put widows and orphans front and center. Of some 6,000 respondents, almost half had bought no more than two bonds (each costing about $5,000 at their time of issue). Two-thirds were over the age of sixty and the same proportion indicated they had bought the bonds for retirement income. Twenty-one percent indicated the WPPSS purchases were the first bonds they had ever bought – and 87 percent said they "would never again buy a bond issued in the state of Washington."[73]

The bondholders' ire was evident in the individual lawsuits consolidated into MDL 551. In some instances, their anger extended to the lead plaintiff in that suit, Chemical Bank, the bond trustee and the nation's sixth-largest bank. Although allied in their efforts to get Northwesterners to meet their obligations, some investors doubted that Chemical (and the investment bankers who marketed the bonds) had their interests at heart. A decade and a half after default, C. Richard Lehmann, an investor who was active in the emerging bondholders' movement, still heatedly maintained that it was "outrageous" that the bondholders should be pictured as "speculators, fat cats . . . [A]ctually the real culprits here were the underwriters. . . . They're the ones who should bear the onus." As for the bond trustee, Lehmann stated that Chemical Bank "mishandled the thing from the beginning." Their ties to the underwriters and other securities firms counted more for the bank than their prospects of recovery for the bondholders. "Morality," observed Lehmann, "is very low on Wall Street."[74]

Despite the conflicts, Chemical attempted to rally support for its legal strategy, calling meetings of bondholders in New York, Chicago, and

[72] For figures on UIT holdings, Zigas, "Unit Trust Investors Bought 16%," *The Bond Buyer*, November 1, 1983, 1; Nuveen sales reported in "Nuveen Trusts Sell All Holdings of WPPSS Bonds," *The Bond Buyer*, November 8, 1983, 7; institutional investors listed in *Clearing Up*, June 24, 1983, 8. Needless to say, these calculations are speculative. More importantly, they refer to a hypothetical "average" investor without information about the size distribution of the bond holdings. Median family income from *The American Almanac 1996–1997: Statistical Abstract of the United States*, 116th edition (Austin, TX: Hoover's, 1996), Table 718, 466.

[73] *Clearing Up*, November 22, 1985, 12–13.

[74] Lehmann interview.

Seattle for October 4, to be linked together by closed-circuit television. The bank publicized the gatherings with full-page notices in the *Wall Street Journal* and other papers and announced its purpose as political: "The only reason we're doing this is for political action and to allow bond-holders to come to the forefront and tell their stories to the public and leg-islators." Nearly 8,000 bondholders had responded to the advertisements; they owned about one-third of the defaulted bonds. However, total atten-dance disappointed organizers – 1,400 at the Felt Forum of New York's Madison Square Garden, 200 in Chicago, where the bank had planned to accommodate 2,000, and 400 in Seattle. *Clearing Up* reported that the Seattle audience "listened in cold silence" to Chemical Vice President Berls and three attorneys representing the bank.[75]

Berls had already encountered the question posed most frequently at the meetings: Why had Chemical not sued the underwriters and oth-ers involved in the financing of the terminated projects? He had replied that "the underwriters never made any misrepresentations to Chemical," which was no doubt literally true since Chemical never owned any of the disputed bonds itself. At the Felt Forum, the panelists pointed out that the bank had focused attention on those who had undertaken the projects and had promised to pay for them. Whether the response made good legal sense or not, it apparently left many of the assembled bondholders dis-satisfied.

Gloom and One Bright Spot

By mid-1983, with four of its five nuclear construction projects halted, the Washington Public Power Supply System was an acutely troubled organization. Mike Leddick, project manager for the suspended WNP-3, had been a candidate for Managing Director that spring but had lost out to Donald Mazur. He complained to a reporter, not about Mazur's selection but about the shutdown of his project. "I'm really upset about what's happening... If the average person knew what they were frittering away they would do something different... There is such an information gap between perception and reality." A few weeks later, he quit. Leddick was only one of sixty-eight Supply System employees who resigned between May and July. Other top managers were among those departing. Project managers for WNP-1 and 2 had left earlier in the year. Executive Board

---

[75] *Clearing Up*, August 19, 1983, 10.

members Durwood Hill and C. Michael Berry resigned their positions that summer. Berry had quit the Board before, in 1982, but had later accepted reappointment. He had been a questioning voice. In May he had told a Tacoma newspaper, "Washington voters should have long ago put a lid on the amount of money WPPSS could spend."[76]

Besides its internal disruptions, the Supply System remained under tremendous pressure about WNP-3. The four investor-owned utilities with shares in the project were seeking an order from Judge Bilby to restart construction. The estimated cost to complete the project was nearly a billion dollars, but with the bond market closed to the Supply System there was no obvious source of these funds. Moreover, for technical reasons, if a restart were delayed, the costs of gearing up again would mount. Deputy Managing Director Alexander Squire estimated that it would be very difficult to preserve readiness on the Satsop site and recommence construction after January 1, 1984.[77]

One slender hope for restarting WNP-3 came from Washington, D.C., Senator James McClure (R-Idaho) introduced a rider to an appropriations bill to allow Bonneville to finance the completion of the project through a new nonprofit corporation to be established in the state of Washington. "There is a great deal of concern that the inability of WPPSS to go to the bond market will cause a general collapse," McClure warned. Pushed by the private utilities with stakes in the plant, McClure's bill initially drew backing from Scoop Jackson and Oregon's moderate Republican Mark Hatfield. Ohio Democratic Senator Howard Metzenbaum, however, proved to be an implacable opponent of the plan, which he called a scheme of right-wing Republicans "asking the government to bail out business." Public utilities were also suspicious, referring to it as a "mystery financing plan."[78] Within a few weeks, McClure's proposal was dead. Senator Jackson passed away on September 1, 1983, removing the region's most powerful politician on energy and utility issues. A House subcommittee hearing on the proposal in mid-September found little support.

---

[76] Wanda Briggs, "Mothballing No. 3 Isn't Fair, Says Its Manager," *Tri-City Herald*, June 4, 1983; Bob Lane, "Uncertainty causes skilled WPPSS employees to leave," *Seattle Times*, August 15, 1983; *Clearing Up*, July 1, 1983, 2; Berry statement in *Tacoma News-Tribune*, May 10, 1983; resignation noted in *Clearing Up*, August 26, 1983, Late Break, no page number.

[77] *Clearing Up*, August 26, 1983, 4, 8.

[78] Ross Anderson, "Plan to save N-plants takes House by surprise," *Seattle Times*, July 20, 1983; "Senate Debates WPPSS Aid Bill; Bank Sues on 4,5," *The Bond Buyer*, August 4, 1983, 1.

Hatfield, citing a "firestorm" of opposition, reversed his position. On September 19, McClure withdrew his rider, although he announced that he might reintroduce it as part of a more comprehensive program to deal with the Supply System's financial plight.[79]

That program, he and others hoped, could come from the report that Charles Luce's Governors' Advisory Panel was preparing on solutions to the major problems facing the Supply System. Originally intended for release in October, the study appeared shortly before Thanksgiving 1983. It was already weighted with heavy political expectations on all sides. Its main proposal was that a new federally chartered regional corporation take over the Supply System. Using funds from a regional rate increase, WNP-3 could be completed. Bondholders on the terminated projects would get a payout of some 36 cents on the dollar, with those who bought up bonds at depressed prices after termination excluded from the deal to avoid the political embarrassment of speculative windfalls going to those who had bought the bonds for as little as twelve cents on the dollar. Congressional sources made it clear that without this differentiation, the necessary federal enabling legislation would be doomed.[80]

The response to the Luce Panel report was a bi-coastal jeer. "It's absolutely preposterous and an insult," an East Coast source reflecting bondholders' interests told *Clearing Up*, while on the Pacific a source sneered, "They're trying to fix something that isn't broken." A month later, *Seattle Times* columnist Shelby Scates pronounced a mordant epitaph. The report was like James Joyce's *Ulysses*, "much discussed, little read and damned near as heavy."[81] Ironically, years later, a leader of the WPPSS bondholders gave a more favorable verdict to Luce's proposal. Luce's analysis, C. Richard Lehmann stated, was entirely accurate. It "would have been cheaper for everybody concerned if they had listened to him."[82]

---

[79] Frank Gresock, "McClure to Rescind His Measure to Aid WPPSS 1,2,3," *The Bond Buyer*, September 19, 1983, 1.

[80] The question of how to deal with post-termination bond purchasers echoes the debate about paying off Revolutionary War debt. In 1790, when Alexander Hamilton proposed in his Report on Public Credit that the new federal government fund the Continental debt and assume the remaining states' debts, James Madison countered that payments should be graduated to reflect the fact that speculators had bought these obligations at very low prices and thus stood to make large, unjustified profits. Hamilton thought of graduation as procedurally impossible and unjust to those who had taken the risk of buying debt when its repayment was by no means assured. Hamilton's policies were enacted as part of a compromise that located the new nation's capital between Virginia and Maryland.

[81] *Clearing Up*, November 18, 1983, 1; *Clearing Up*, December 16, 1983, 6, quoting Shelby Scates.

[82] Lehmann interview.

Lehmann's hindsight benefited from the experience of waiting
early 1990s for negotiated settlements. However, in late 1983
report was "stillborn" in the words of one congressional staf.
Congressman Weaver's judgment in March 1984 was pungent: "a mound
of steaming something or other that no one wants to touch."[83] The almost
unanimous lack of enthusiasm for Luce's proposal in effect doomed any
congressional action on the costs of the terminated plants.

Amidst its problems, the Washington Public Power Supply System could
point to one bright spot. Project Two, the only one of the five plants
still being built, was nearing completion. Robert Ferguson's managerial
reforms had pulled construction out of its quagmire. By late 1983, the
Supply System had earned its operating license from the Nuclear Regulatory Commission and on Christmas Day began loading fuel in the plant's
reactor. Commercial operation, previously scheduled for February 1984,
was now anticipated for July. Harold Denton, in charge of the NRC's
reactor licensing, told the press that he was "very pleased" with the work
on WNP-2. Closer to home, Cyrus Noë, often a sharp critic of the Supply
System, called the project's completion a "remarkable achievement."
Despite the approbation and a good deal of self-congratulation, a series
of minor complications forced back the date that the Supply System's first
reactor of its own began to produce electricity for sale until December 13,
1984.[84]

When WPPSS Nuclear Plant 2 went on line, the figures pointed to some
of the woes that nuclear construction had been facing. Originally estimated at less than $400 million, WNP-2 ended up costing about $3.2 billion. Construction had begun in August 1972 with completion expected
in September 1977; it took an additional seven years and three months
to finish, over twelve years in all. Average total cost was calculated at 6.2
cents per kilowatt-hour, nearly three times Bonneville's wholesale preference rate for public utilities.[85] In the week that fuel loading for WNP-2
began, Cyrus Noë mused on the "great power generating station era that

[83] *Clearing Up*, February 3, 1984, 17; Weaver comment in ibid., March 16, 1984, 9–10.
[84] Miller, *Energy Northwest*, chapter 21 ("WNP-2 Run-up to Operations") is a very good
narrative; Noë's praise quoted on p. 458. Denton comment in Joel Connelly, "First completed WPPSS N–plant gets high marks," *Seattle Post-Intelligencer*, March 2, 1984.
[85] Joel Connelly, "WPPSS 2 plant gets its operating license," *Seattle Post-Intelligencer*,
December 21, 1983; Carrie Dolan, "WPPSS Gets License to Run Nuclear Plant," *Wall
Street Journal*, December 21, 1983. The cost per kilowatt-hour estimate comes from
Clearing Up, June 1, 1984, 2. Admittedly, Northwest energy planners had never claimed
that nuclear power would be as cheap as hydro, but at 6.2 cents per kilowatt-hour,
WNP-2 would produce very expensive electricity.

began in the 1930s." That era might not be over, but "we are...quite obviously on the downslope." He continued:

The future of utilities is less in building large scale generating plants and more in small scale projects, conservation, diversification, joint operations, regional planning and multi-regional planning... [C]reating larger and larger supplies of electricity is not a prerequisite for economic growth. Not any more. That lesson is not bad news for utilities, it is good news.[86]

No radical, and careful to avoid bias in his reportage on the Supply System, the editor's statement signified that the Northwest energy community's consensus for construction had vanished. Heterodox ideas were entering the mainstream. A few months later, Peter Johnson spoke to the Seattle Rotary Club and warned that electrical utilities "have become servants of giant construction projects rather than servants of their ratepayers." Bonneville had to stop "chasing construction projects," or it would become a western version of LILCO, the utility responsible for New York's ill-fated Shoreham project. Bonneville's leadership had come a long way from the days of Donald Hodel's 1975 attack on "prophets of shortage."[87]

In a playful mood in spring 1984, Noë asked readers to propose a nickname for the new nuclear plant. The suggestions indicated more than a little cynicism: Young Frankenstein II, Hodel's Folly, Spectacost II, Faulty Towers, Hanfordsaurus Rex. The winner, he proclaimed, was Moby Deuce. Twelve years earlier, WNP-2 had looked like a rational and economical undertaking; by 1984, it seemed more to be a costly, self-destructive obsession.[88]

MDL 551 – The Early Stages

Judge Richard Bilby assumed the task of managing the complexities of Multidistrict Litigation 551 with visible anxiety. In late summer 1983 he told a crowded courtroom that the parties "are asking me to become the energy czar of the Northwest, and that scares me. You people have been wrestling with this problem since the early '70s and not very successfully." He intended to resolve which parties would remain in the suit and appoint lead counsel for the case by the end of 1983. "Then we will be ready to roll, listing witnesses and taking depositions. It will be a lot better organized when we really get the players into the slots and we know who they are."

[86] *Clearing Up*, December 30, 1983, 8–9.
[87] Johnson quoted in *Clearing Up*, May 18, 1984, 2.
[88] *Clearing Up*, June 8, 1984, 8.

The trial itself would start in approximately two years, he predicted. His forecast proved to be three years too optimistic.[89]

In the following weeks, Bilby attempted to bring some order to the case. He formally joined the bondholders' actions together with Chemical Bank's suit after the default. In December, saying that he was concerned about "shotgun allegations" against individuals only peripherally involved, he dismissed charges against over 400 members of participating utility boards, their lawyers, and advisors. He initiated discovery procedures for the millions of pages of documents that the case would require. He took on other problems, large and small. One of the latter was how to refer to the agency at the center of the dispute. The pun on the acronym might prejudice jurors. When one lawyer referred to "WPPSS," Bilby corrected him: "The newspapers can call it what they want, but we'll call it the Supply System."[90]

Early activity in MDL 551 ran concurrently with appeals of the state Supreme Court's June 15 Chemical Bank case ruling. However, the legal issues involved in the two cases were very different. In Chemical Bank, the bond trustee had been trying to affirm the validity of the Participants' Agreements that the eighty-eight utilities had signed in 1976. Had the bank succeeded (and had cases with the same question at stake in Oregon and Idaho gone the same way), the Supply System would willingly have passed on payments from the Participants to the bondholders. In MDL 551, the issue was not whether the Participants' Agreements were valid. Rather, both Chemical and the class action bondholder plaintiffs charged a broad spectrum of defendants with violations of federal, state, and common law for misrepresentations and omissions in the marketing of the bonds. Indeed, the fact that the Participants had successfully negated the seemingly airtight take-or-pay clauses in the Agreements was an important part of the plaintiffs' case in MDL 551. They contended it showed that earlier promises of the bonds' security had been misrepresentations. In other words, MDL 551 took the overturning of the Participants' Agreements as a fait accompli. That this was a case about misrepresentation

[89] "Late Break One," *Clearing Up*, September 2, 1983, cover page; Martin Heerwald, "Judge outlines procedures on WPPSS litigation," U.P.I. Regional News, September 1, 1983.

[90] Martin Heerwald, "Judge issues who's in, who's out order in WPPSS case," U.P.I. Regional News, December 9, 1983; "Dismissal Moves are Rejected in WPPSS Suit," *The Bond Buyer*, December 12, 1983, 3 (The headline refers to Bilby's refusal to dismiss certain other defendants.); U.P.I. Regional News, September 3, 1983; Heerwald, "Judge outlines procedures..."

rather than contract may have seemed a technicality to aggrieved bond-holders, but it shaped the way that MDL 551 evolved and wound toward eventual resolution.

Multidistrict Litigation 551 was enormous by almost any measure – the money at stake, the number of parties involved, the platoons of lawyers who took part, the estimated 140 million pages of documents to be pro-duced, the 300 witnesses deposed, and the 4,000 entries in the court docket all demonstrated its magnitude.[91] The case was extremely difficult, in part due to the legal complexities that came with its size, in part too because the case posed major problems and threats for each of the major parties involved.

For the bondholders, the hurdles were threefold. In the first place, the legal challenge was daunting. Historically, regulation of the sale of munic-ipal securities had been far less stringent than sales of corporate issues. The abuses that provoked the Securities Act of 1933 and the Securities Exchange Act of 1934 had been in the private sector, and the demand for remedies was greater there. The banks and insurance companies that had long been the main purchasers of municipal bonds were less vulner-able to misrepresentation than individual investors. Moreover, municipal securities issuers were frequently small communities and districts. Bur-dening them with excessive legal expenses to borrow modest sums could block their entry to the bond market. Concern for principles of states' rights provided a constitutional rationale for lighter regulation of state and local issues at the federal level.[92] Yet, although registration proce-dures under the New Deal legislation exempted municipal securities, the antifraud provisions of the Securities Exchange Act did apply to them.

Under a key provision of the Securities Exchange Act, Section 10b, showing that the Supply System had misstated or omitted information would not suffice for the plaintiffs in MDL 551. They would have to demonstrate that the misrepresentations were material – that they would have affected or at least have been a consideration for a reasonable per-son deciding whether to purchase the bonds – and, more importantly, that they were made either recklessly or with an intent to deceive. This latter

[91] The staggering figure of 140 million pages of documents is repeated in several sources. See, for instance, William Horne, "Chemical Bank et al. v. WPPSS et al.," *The American Lawyer*, March 1989.

[92] Roger L. Davis and Reece Bader, "SEC Enforces Municipal Disclosure Obligations," *National Law Journal*, August 5, 1996, B4 provides a useful overview of the development of municipal securities regulation. See also Lamb and Rappaport, *Municipal Bonds*, 225–244.

provision, known as *scienter*, made proof of fraud in the sale of municipal bonds very difficult. It also implied that the pretrial discovery and deposition processes would have to delve deeply into the history of the failed projects not only to determine whether claims in the bond official statements and elsewhere were false, but also what the intentions of their makers had been. Throughout MDL 551 and related cases, bondholders looked for ways to avoid the *scienter* burden, but they could not escape it.

When he approved the out-of-court settlements that finally ended MDL 551, Judge William Browning, who had replaced Richard Bilby on the case, evaluated the settlements according to the strengths and weaknesses of the plaintiffs' case. Without predicting a hypothetical verdict, he clearly indicated the legal difficulties *scienter* posed for Chemical Bank and the bondholders. The plaintiffs, he noted:

needed to show that hundreds of officers and employees of a large number of public and private entities conspired over a period in excess of ten years, frequently in public meetings, to borrow money they knew they could not and would not repay to build Projects that would be too costly to complete and were not needed in any event, often contrary to their own individual interests as Bondholders.[93]

A second problem facing the plaintiffs was the tension between bondholders and Chemical Bank. This surfaced almost immediately and lasted even beyond the final settlements. For many investors, Chemical was too closely allied with the investment bankers and brokerage houses that had been selling the bonds all along. Bondholder activist Richard Lehmann charged that the bank's "ongoing business relationships were more important to them than what could be recovered out of the situation."[94] Several incidents heightened investor suspicions of the bank. At one point, Chemical was grudgingly supporting a suit in Washington state court against the state auditor; at the same time, however, the bank became managing underwriter for a large Washington state bond issue. According to the lead plaintiff in the state suit, Chemical's apparent conflict of interests was "a horrendous thing.... It's unethical."[95]

Chemical's refusal to include Wall Street interests in MDL 551's otherwise exhaustive defendants' roster had also continually rankled some bondholders. One of the trustee's attorneys, Michael Mines, invoked a

---

[93] In Re Washington Public Power Supply System Securities Litigation, 720 F. Supp. 1379 (Arizona, 1989), 1389. Hereafter cited as MDL 551.

[94] Lehmann interview.

[95] George Yacik, "Chemical Role In Wash. Issue Angers WPPSS Bondholders," *The Bond Buyer*, October 11, 1984, 1.

notion of equity in his account of Chemical's thinking about whom to sue:
"Even the littlest guy in town is responsible if he helped approve the bond
sale. We didn't want to just name the wealthy participants. We wanted to
be fair."[96] This rationale could hardly have pleased bondholders, large or
small, hoping to recover as much of their investment as possible. When a
different bondholder group filed another suit in state court emphasizing
the culpability of the "professional" defendants who marketed the bonds,
Chemical refused to pay legal expenses and stated that it "views the filing
of a new complaint by a separate group of bondholders as lacking any
real significance."[97]

A third problem for the plaintiffs in MDL 551 was the fact that real
interests as well as differing legal and political evaluations divided the
bondholders themselves. If WPPSS and others had misrepresented their
bonds, which buyers had suffered? Those who sought high returns during
the years when WNP-4 and 5 bonds were cascading regularly onto the
municipal market were the obvious losers. But what of those who purchased between January 22, 1982, the date the projects were terminated,
and the June 15, 1983, decision invalidating the Participants' Agreements?
Or those who speculated in deeply discounted bonds after the June 15
verdict and subsequent default? In MDL 551, the bondholder plaintiffs
were originally limited to those who bought prior to termination. Later,
purchasers between termination and the Washington Supreme Court's
decision were included, in a separate category. Those who had bought
their bonds after June 15, 1983, were excluded. These late investors,
who knew that their bonds would not be paid off without a reversal
of the Chemical decision, had bought the bonds at deep discounts and
had to hope either that an appeal of that suit or that individual securities law suits would provide them a windfall for their low-priced purchases. By 1989, they owned about twenty percent of the WNP-4 and
5 bonds.[98]

The divisions were evident by January 1984. *Clearing Up* reported
that the Chicago group was charging that a newly formed National 4/5
Bondholders' Committee was "dominated by speculators" and "thus not

---

[96] Carrie Dolan, "WPPSS, Others Sued by Chemical, Which Charges Negligence," *Wall
Street Journal*, August 4, 1983.

[97] *Clearing Up*, June 1, 1984, 10–11.

[98] MDL 551 at 1403 cites the 20 percent figure. Although *Clearing Up* reported that "the
bottom has dropped out" of the market for Project 4 and 5 bonds in the wake of the U.S.
Supreme Court's refusal to take the Chemical Bank case on appeal, trading did continue
sporadically. *Clearing Up*, May 10, 1985, 2.

proper people to speak for bondholders." Cyrus Noë pointed out that all of the leading figures on the National Committee outside the Northwest had bought their bonds since termination. "The tension between late buyers and those who bought when the 14 issues came out is very considerable." When the National Bondholders' Committee met in Dallas in January 1984 and elected a bond broker who admitted buying his bonds after termination as the group's chair, the Northwest regional committee withdrew: "We're not going to get to first base with Congress with a speculator as committee chair. And on top of it all, he's...probably sold WPPSS 4/5s in the first place. What standing can he possibly have to lead a parade of widows and orphans?"[99] Late buyers knew their chances of winning a suit like MDL 551, based on misrepresentation, were poor. They had purchased after the WPPSS crisis had become common knowledge. The late buyers thought their best chance would be to pursue claims based on breach of the bond contracts and their promises to pay, come hell or high water.[100]

The stakes for defendants were also extremely high and their tasks formidable. The dimensions of the threat varied. The Supply System itself was, oddly, quite invulnerable to financial penalties in the case. Although Chemical and other plaintiffs wanted very much to breach the "Chinese Wall" that kept them from the assets of net-billed projects, as the case developed, WNP-1, 2, and 3 were effectively insulated from the MDL 551 claims.[101] Nevertheless, the Supply System's position was precarious. It had consistently maintained the validity of the Participants' Agreements and had been shocked by the refusal of the utilities to pay for the terminated plants. To be held guilty of misrepresenting the bonds it issued would be to say that the legal and energy planning for WNP-4 and 5 had been not only inept but dishonest. It would doubtless be a ruinous political blow to an agency that already had a bad reputation.

For the eighty-eight utility Participants, a loss in MDL 551 threatened to reimpose the financial burden that they had escaped by winning the 1983 Chemical Bank case. Rates would soar, demand would likely fall sharply, and utilities might find themselves in the death spiral they thought they had escaped when the Washington Supreme Court invalidated the Participants' Agreements. Some of the utilities had liability insurance, but in most cases, it would not cover any judgment requiring them to pay the full principal and interest on the defaulted bonds. In addition,

---

[99] *Clearing Up*, January 20, 1984, 14.   [100] *Clearing Up*, January 6, 1984, 10–11.
[101] MDL 551 at 1401 discusses the legal barriers to collecting from the Supply System.

a judgment for the full $2.25 billion could leave the utilities liable for huge legal bills as well. Thus, MDL 551 might sever the lifeline that the Chemical Bank decision had thrown the utility participants.[102] Financially well-heeled utilities were especially threatened. Under the legal doctrine of joint and several liability, a utility with "deep pockets" might have to dig out large sums to pay for judgments that other defendants were unable to meet.

Bonneville's situation was distinct but also serious. Throughout the Supply System's travails, Bonneville asserted (tendentiously, as chapter 3 shows) that it had not been actively involved in the inception of Projects 4 and 5. Moreover, under common law and the constitutional principle of sovereign immunity, the federal agency could not be sued unless it waived its exemption. Under the Federal Tort Claims Act, the process of suing Bonneville was convoluted. Each individual bondholder first had to file an administrative claim against the BPA. Some 13,000 investors had requested inclusion in this action by early March 1984, representing well over half the total value of the bonds. Judge Bilby then ruled that the federal agency could be included as a defendant in MDL 551. Indeed, one group of plaintiffs maintained that Bonneville was the real villain of the story. Bonneville was the prime promoter of the projects and was negligent because it "should have known" that Supply System personnel "were incapable of planning and managing projects of the size and scope" of WNP-4 and 5.[103]

Bonneville bounced on and off the defendants' roster during the long pretrial period. In early 1986, Judge Browning ruled that Bonneville was immune from the plaintiffs' claims for its involvement in the projects. (Under the Constitution and federal law, if a federal agency had discretionary power to take a particular course of action, it cannot be sued for that action. Suits are permissible only if it acts beyond the scope of its discretionary authority.) He then permitted the plaintiffs to modify their charges to assert that BPA had acted in areas beyond the agency's authority, but in fall of 1987 dismissed these claims as well. However, as long as BPA was a defendant, Bonneville was "the deepest pocket of all." A judgment against it could be extremely costly. In response to the

---

[102] Jay Mathews, "Pacific Northwest May Face a Long Rainy Day if 'Whoops' Bust," *Washington Post*, December 29, 1982. Judge Browning made the point about joint and several liability in MDL 551 at 1409.

[103] *Clearing Up*, August 31, 1984, 11.

threat, the U.S. Department of Justice in late 1983 established a special legal office in Portland to handle WPPSS-related litigation on behalf of the BPA.[104]

Another group with much to lose from a plaintiffs' victory in the MDL 551 case was collectively known as the "professional defendants." These ranged from the projects' architect-engineering firms and the engineering consultant firm through the Supply System's law and accounting firms, its bond counsel, and its financial advisor during the period of WNP-4 and 5 bond sales, to the lead underwriters, and the bond rating agencies Standard & Poor's and Moody's. As noted before, Chemical had not sued these organizations, but the bondholders in the class action had named them. MDL 551 threatened the professional defendants' finances and their reputations. The underwriter defendants were wealthy and powerful organizations consistently at the forefront of marketing all securities, not just municipal bonds. The bond counsel, Wood Dawson, was a venerable New York firm with an important role in the bond market. The Supply System's special counsel, Houghton Cluck Coughlin & Riley, had been a linchpin of the public utilities movement in the Pacific Northwest and occupied a respected position in the Seattle bar. R.W. Beck, the engineering consultants who served countless Northwest utility projects, were also stalwarts of many decades. United Engineers and Constructors and EBASCO Services were two of only six American firms that served as architect/engineers for nuclear projects.

Individual defendants, usually lacking personal liability insurance to protect them against legal claims, also had to view MDL 551 as a danger. Although Supply System board members and those who served on the Participants' Committee were not rich enough to make a substantial contribution to a judgment, these defendants had their personal reputations and careers to defend.

The perils of MDL 551 gave parties on both sides reason to look for a settlement. The pretrial process seemed to produce little clarity. Legal expenses ran high. The trial itself would be a gamble. Expected to last as long as a year, featuring immense amounts of confusing and technical testimony and evidence, this was a trial whose outcome could hardly be predicted. A jury could return a verdict requiring full payment or none at all and could allocate financial burdens in an infinite variety of ways.

[104] Rich Arthurs, "Justice Steps Up WPPSS Defense Effort," *Legal Times*, February 13, 1984, 1.

Many heads must have nodded when Peter Johnson, in a retirement speech in 1986, proclaimed, "God, we've *got* to settle [WNP-] four and five."[105]

The Path Toward Trial

Judge Bilby had estimated that trial preparations in MDL 551 would take about two years. In January 1984, he optimistically moved his predicted starting date up to June 1985 and indicated that the trial should take about three months. In fact, in June 1985, two years after the Washington State Supreme Court had ruled the Participants' Agreements invalid, Cyrus Noë commented, " . . . we must conclude that the MDL 551 action is a mess." Almost nothing had been resolved.[106] By that point, Judge Bilby himself was no longer on the scene. It would take over three more years before the case came to trial.

Everything about the MDL 551 case was mammoth. It was obvious from the start, for example, that it was going to call forth vast amounts of paper. In spring 1984 discovery began. The process, in which opposing parties were required to provide information to each other, soon became a bone of contention. How much would they have to bring forth, and on what topics? What time period would discovery cover? Under Federal Rules of Civil Procedure, each party is entitled to demand a great deal from its adversaries. *Clearing Up* reported estimates that discovery would generate 100 million pages of paper. A few weeks later, defendants claimed there would be 140 million. Plaintiffs' attorneys contended that the defendants not only had to produce documents going back a decade, to the start of planning for WNP-4 and 5, but earlier – perhaps as far back as the early 1960s. The judge pointed out that this would delay the start of depositions of witnesses, but he allowed the plaintiffs to get at some material related to the net-billed plants. At one point, noting that WPPSS had pled a defense that it had relied on their bond counsel, the plaintiffs argued that this meant that even the System's communications with its law firms should be handed over for discovery.[107]

Preparing to depose witnesses also was a complicated process. Estimates of the number of needed depositions ranged from three hundred to five hundred. Eager to start, bondholders' attorneys had asked to begin

[105] *Clearing Up*, July 11, 1986, 2.
[106] *Clearing Up*, January 27, 1984, 15; June 21, 1985, 8.
[107] Disputes over discovery are noted in *Clearing Up*, March 9, 1984, 11; March 30, 1984, 10; June 29, 1984, 3; July 6, 1984, 1; August 3, 1984, 1.

the procedure by June 1984, but Judge Bilby announced that taking the depositions would be organized in four overlapping tracks. The first group was to begin September 10, 1984. The flood of discovery requests made a fall start impossible, and eventually the initial deposition sessions began in the first week of 1985. Looming over them was a threat from Bonneville. Since the agency wanted to avoid being included as a defendant in MDL 551, the agency asserted a right to take depositions from all 15,000 bondholders who had filed claims against it. Eventually, the BPA argument failed, but it indicated MDL 551's potential to spiral out of all control.[108]

In 1993, as Supply System litigation crept toward conclusion, James Perko, the System's Chief Financial Officer, reminisced about his own deposition. Grateful that the Board had agreed to hire "great lawyers... They're expensive, but they're worth every dime," Perko recalled the process as grueling. Following a month's preparation, he underwent three months of deposition. His "conservative" estimate of the cost of his own deposition was between three-quarters and one million dollars.[109]

On January 21, 1985, MDL 551 hit a major roadblock. Bilby announced that he was stepping down from all WPPSS-related litigation. He had just found out that his father and stepmother owned WNP-3 bonds with a face value of about $100,000. William D. Browning, also a Federal District Judge from Tucson, immediately replaced him. Judge Browning, a recent Reagan appointee, had only about nine months of judicial experience when he found himself in charge of MDL 551 and related cases. Judge Bilby took the unusual step of reassigning his law clerk to Judge Browning, but the new jurist had little background in securities law and would in any event need time to study the case. Bilby's recusal threw into question the status of the decisions he had already rendered. Judge Browning said that his substitution should not significantly slow down progress on the cases, but observers fretted.[110]

While the parties and their attorneys tried to digest the sudden judicial replacement, those who had been unhappy with Bilby's decisions saw an

[108] *Clearing Up*, March 30, 1984, 10; "BPA Made WPPSS Defendant; Cost Estimate Cut on Units 1, 3," *The Bond Buyer*, April 30, 1984, 1; *Clearing Up*, September 14, 1984, 1; February 22, 1985, 10–11.

[109] Howard D. Sitzer, Cyrus Noë, and James D. Perko, "The Washington Public Power Supply System: Then and Now," *Municipal Finance Journal*, 14, 4 (Winter 1994): 76–77.

[110] Howard Gleckman, "Judge Quits; Cites Parents' Holdings of Project 3 Bonds," *The Bond Buyer*, January 23, 1985, 1; *Clearing Up*, February 1, 1985, 10–11.

opportunity to have them vacated. They asked Browning to discard all of Bilby's rulings on the grounds that the conflict of interest had tainted them. Ruling in May 1985 on this so-called "square one" motion, Browning agreed that the earlier decisions had to go. "[J]udicial economy and efficiency are questions that must be subordinated to the paramount questions of public confidence in the judicial system," he wrote. "Two years of work down the drain," one lawyer complained.[111]

Judge Browning then directed attorneys to bring "one-line" requests to him to reinstate the motions they had presented to Judge Bilby. What this would mean was ambiguous, and some lawyers maintained that the procedure was unworkable. This was when Cyrus Noë abandoned his usual cool tone and complained "that the MDL 551 action is a mess. The major orders by Judge Bilby are vacated. The procedure to renew them with the court is a vague and unsystematic invitation by the judge to throw one-liners in the hopper along with some new case law lists and let the judge play Trivial Pursuit with the results."[112]

Another pursuit was anything but trivial. This was the search for a trial venue. Parties from outside the Northwest did not want the case tried in Seattle. Securities underwriters had presented survey results indicating that an impartial jury could not be empaneled there. The underwriters took the matter very seriously. Their motion for a change of venue weighed twelve and a half pounds. The utility defendants, Bonneville and the Supply System responded with their own survey purportedly demonstrating the opposite. On September 7, 1984, Judge Bilby conducted a mock *voir dire*, an examination of jurors, with Washington residents that persuaded him that many of them were highly aware of the WPPSS situation and of their interests as ratepayers. However, Northwestern defendants and their counsel were also enthusiastic about the *voir dire* experiment, hoping that it would convince the judge that there were enough open-minded citizens in the area to merit keeping the trial in Seattle. Apparently the mock *voir dire* aided those seeking a change of venue. More than two-thirds of potential jurors answering a questionnaire were disqualified. Bilby brought in the remaining prospects and asked them if they believed that a judgment against the utilities would raise their electric bills. All raised their hands.[113]

---

[111] Howard Gleckman, "Judge's Negation of Prior Rulings Puts WPPSS Back at Square One," *The Bond Buyer*, May 24, 1985, 1.

[112] *Clearing Up*, June 21, 1985, 8.

[113] *Clearing Up*, May 18, 1984, 3; Rich Arthurs, "Mock Voir Dire Used to Assist in Venue Decision," *Legal Times*, September 17, 1984, 1.

In October 1984, Judge Bilby ruled that the trial would be moved out of the Seattle area. He did not say where it would go, but later that fall reports indicated he was thinking of using a converted junior high school auditorium in San Diego. Cyrus Noë complained that San Diego would be "an expensive mistake." No doubt the quarrelsome atmosphere surrounding MDL 551 contributed to an outburst from the judge in December: "Someone said to me the other day that it's impossible to try this case. They told the wrong person that. We'll try it all right. It's never been done, but there has to be a first fish, and I'm willing. We'll try it."[114]

William Browning's assumption of Bilby's role in MDL 551 reopened the venue question. At hearings in 1986, Browning said that he didn't doubt the capacity of Washington citizens to be fair and impartial, "but it's almost too much to expect of them." In February 1987 the judge announced that the trial would take place in Tucson. Even then utility attorney Albert Malanca appealed to the Ninth Circuit Court of Appeals to return the case to Seattle. People in the region could be fair, he maintained. "Greater Puget Sound has half a million people. You can't tell me that all of them have a close friend or relative that buys public power from WPPSS."[115] The appellate court rejected the plea. By 1987, the trial date had slipped back to September 1988, more than five years after the default.

The size of MDL 551 meant that Tucson had no courtroom large enough to hold the trial. Indeed, no federal courtroom in the country could have accommodated a trial with over 100 defendants and as many as 300 lawyers – not to mention the tens of thousands of individual plaintiffs. Judge Browning and the attorneys agreed to have a hotel ballroom at Tucson's Ramada Inn Downtown converted to a facility where the trial could take place. California architect Michael Ross was chosen to design the Tucson room; he had done two other "mega-courtrooms" for class action cases, but this would be his largest project. He planned thirty lawyers' tables, each linked to a computer database. "It's kind of like a large state legislative house," he explained. The hotel's general manager anticipated "tremendous" sales of food and drink to those who came to the courtroom.[116]

---

[114] *Clearing Up*, December 14, 1984, 15–16.
[115] *Clearing Up*, April 5, 1985, 8; "Judge Wants WPPSS Trial to Be Moved," *Eugene Register-Guard*, April 22, 1986; "Tucson Is Named as Site for Trial of Issues Involving WPPSS 4, 5 Bonds," *The Bond Buyer*, February 13, 1987, 4; Vicky Stamas, "Hearing Set Thursday on Site choice for '88 Trial in WPPSS Bond Default," *The Bond Buyer*, October 5, 1987, 5.
[116] "Ballroom Converted for WPPSS Trial," *Eugene Register-Guard*, July 26, 1987.

The Other Cases

Throughout the preparations for MDL 551, other legal actions shadowed the case. Lawsuits numbered in the dozens, but we can group the most important in three categories. First, there were suits by bondholders pursuing other paths to recover their investments. These included investors who disputed Chemical's strategy in MDL 551 and those excluded from that suit as late purchasers of the bonds. Second, there was the contest over the apportionment of costs between the terminated and the net-billed projects. In this category we can also place a complex dispute in which the region's private utilities, owners of 30 percent of Project 3, objected to that plant's mothballing. Finally, there were high-stakes controversies between WPPSS and its contractors and suppliers.

Suits named after their lead plaintiffs, bondholders Arthur Hoffer and Fredric Haberman, both tried to take advantage of the absence of a *scienter* provision in Washington state securities law. Without this, plaintiffs hoped to win their case by demonstrating negligence instead of the intentional misrepresentation that *scienter* demanded. The Hoffer suit chose to name Washington's state government and its auditor as the main defendants, even though these parties had not been on the long list of defendants in MDL 551.[117] Perhaps for this reason, Cyrus Noë described the Hoffer suit as the legal equivalent of a "Hail Mary" desperation shot in basketball. Not surprisingly, a Seattle trial judge firmly slapped it out of court in 1985. That year, pressed by utility lobbyists, the state legislature added a *scienter* provision to Washington securities legislation. If this was not enough to get rid of bondholder actions, the next year it amended the law to emphasize that the provision should be applied retroactively. Despite this, the Washington Supreme Court overturned the dismissal of the Hoffer suit and returned the case to trial court. However, as we shall see, resolution of MDL 551 in 1988 eventually denied the Hoffer plaintiffs their day in state court.[118]

A comparable fate befell the Haberman suit, filed in May 1984. The plaintiffs here wanted to go after the professional defendants, especially the underwriters, and hoped that state courts would be a more congenial forum. Dismissed by the trial judge in 1986, the Washington State

---

[117] *Clearing Up*, November 16, 1984, 1, 9–11; James Russell, "Victims of WPPSS Try a Comeback," *Miami Herald*, November 18, 1984.

[118] *Clearing Up*, July 3, 1985, 7–8, 13; Vicky Stamas, "Washington Officials Must Face WPPSS Bondholder Fraud Suit," *The Bond Buyer*, May 13, 1988, 1.

Supreme Court revived the case a year later, but the justices upheld *scienter* as applicable. Haberman's lawyers appealed this to the U.S. Supreme Court, which in 1988 refused to review the case.[119] Together, Hoffer and Haberman indicate how crucial the *scienter* provision was to the WPPSS defendants.

Another set of legal complications resulted from the physical and financial links between the terminated projects and the net-billed plants. Because Projects 1 and 4, at Hanford, and Projects 3 and 5, at Satsop, had been "twinned" on a common site, allocating costs for shared construction expenses and facilities posed substantial challenges. The cost-sharing litigation was perhaps the most tangled of the Supply System's legal engagements. On its face, the legal issue was an arcane dispute about cost accounting, but hundreds of millions of dollars were at stake. The Supply System conceded that the net-billed projects might owe the terminated ones about $400 million, but Chemical Bank's attorneys said liability might actually reach one billion dollars. Judge Bilby had hoped to resolve cost-sharing issues before MDL 551, but this proved impossible. When Judge Browning took up the issue in early 1989, he remarked, "I've backed away from this case like all of you [lawyers]."[120] Perhaps backing away was the wisest course. The sums involved shrank over time. Finally, in 1995, with a retired federal judge handling negotiations, the parties settled for a payment of $55 million to the bondholders.[121]

The mothballing of net-billed project three had also landed in court. As we have seen, the four private utilities owning 30 percent of WNP-3 had objected to the 1983 construction halt. In August, the IOUs asked Judge Bilby for an injunction to force construction to resume. "We're not going to sit around letting WPPSS and the Bonneville Power Administration decide what we're going to do," insisted one utility lawyer.[122] In 1985,

---

[119] The Washington Supreme Court's 1987 decision is at 109 Wash. 2d 107; George Yacik, "Lawyers for Bondholders Find Conflict in WPPSS Rulings," *The Bond Buyer*, October 12, 1987, 1; Geoffrey Campbell, "Supreme Court Declines to Review Fraud Standard in WPPSS Case," *The Bond Buyer*, October 4, 1988, 1. The decision is criticized in Barbara L. Schmidt, "Note: Expanding Seller Liability Under the Securities Act of Washington – Haberman v. WPPSS," 63 Wash. L. Rev. 769 (1988).

[120] *Clearing Up*, February 24, 1984, 1; "WPPSS Judge Promises Ruling in 30 Days on Cost-Sharing Issues Pending Since '87," *The Bond Buyer*, April 14, 1989, 2.

[121] Dennis Walters, "Appeals Court Reverses Lower Court Decision on WPPSS Subsidies of 'Twinned' Plants," *The Bond Buyer*, February 28, 1992, 1; Donald C. Bauder, "Bondholders Settle Suit Over Two Nuclear Plants," *San Diego Union-Tribune*, January 28, 1995; Brad Altman, "Washington State," *The Bond Buyer*, January 31, 1995, 24.

[122] Mark Fury, "Attorney Says WPPSS 3 Action Is Utilities' 'Full-Court Press'," *The Bond Buyer*, August 25, 1983, 3.

negotiators agreed that Bonneville would buy out the private firms' project shares. In payment, the IOUs would receive low-cost power from the BPA. Public power balked at the arrangement, which Congressman Weaver labeled a "raw deal." The settlement did not go into effect until appeals were exhausted in 1989.[123]

A final legal category involved several actions reflecting the difficulties of attempting to construct five large nuclear power plants. These were disputes with contractors and suppliers, some of which posed substantial threats to the financially shaky Supply System. Two of them deserve attention, the first for its financial implications, the second for its bearing on a key legal issue in MDL, 551, whether funds from the net-billed projects could be used to pay obligations relating to the terminated plants.

In January 1985, only weeks after WNP-2 went into operation, WPPSS sued General Electric, the manufacturer of the plant's nuclear steam supply system, over flaws in the containment vessel. The Supply System sought $1.2 billion in damages, a startling amount, but one that the plant's cost escalation to $3.2 billion brought within the realm of reason. Like other WPPSS litigation, the case followed a tortuous path to an unsatisfying conclusion. A jury trial in 1990 ended in deadlock, but the judge castigated GE for its breach of "good faith and fair dealing." Negotiators reached an agreement in 1992, on the eve of a retrial. At GE's insistence and to the dismay of Northwestern utilities, its terms were sealed for three years. When they were revealed in 1995, it turned out that the Supply System had netted only $134.9 million, not in cash but in supplies and services. As historian Gary K. Miller observed, the settlement can only be viewed as a defeat for the Supply System, since the initial damage claims were nearly nine times the settlement amount.[124]

Another controversy, with less money at stake but an important legal issue involved, set the Supply System against steel contractor Pittsburgh – Des Moines Corporation. Even in the company of other WPPSS litigation, the Pittsburgh – Des Moines case stands out for its intricacy, involving claims and counterclaims on construction work on WNP-2, questions of contract interpretation that led to a series of decisions, appeals, and reversals, and a claim by the contractor for unpaid work on WNP-5. The

[123] Howard Gleckman, "Four Investor-Owned Utilities Reach WPPSS 3 Agreement with Bonneville," *The Bond Buyer*, September 11, 1985; a good summary of the case is in Miller, *Energy Northwest*, 436–438.
[124] Larry Lange, "GE Settles WPPSS Lawsuit but Secrecy Angers Utilities," *Seattle Post-Intelligencer*, March 26, 1992; "$135 Million WPPSS Plant Settlement Divulged," *Seattle Post-Intelligencer*, December 20, 1995; Miller, *Energy Northwest*, 518–519.

Supply System maintained that only special funds established to hold moneys for WNP-4 and 5 could be used to pay any judgment against it relating to the terminated WNP-5. Pitt–Des Moines, on the other hand, asserted that money in accounts for the net-billed projects should go towards paying it. In a decision handed down in September 1988, the Ninth Circuit Court ruled in favor of WPPSS on this issue. Pitt–Des Moines could not tap the revenues of the net-billed projects. This verdict, upholding the "Chinese Wall," led Chemical Bank to abandon efforts to obtain money from Projects 1, 2, and 3 to pay off investors in the terminated plants.[125]

The Main Event: MDL 551

In the mid-1980s, hopes for a speedy resolution of Multidistrict Litigation 551 repeatedly evaporated. Judge Bilby's recusal, Browning's "Square One" ruling, imbroglios over the *scienter* standard, discovery and deposition practices, venue and the status of Bonneville Power Administration as a defendant all stretched out the case. Squadrons of lawyers, mountains of papers, and a set of daunting legal issues characterized the maneuvers prior to trial. For one Seattle lawyer, the most telling feature of MDL 551 was the number of attorneys attending depositions. "I've seen depositions that look like law school classes," he remarked. "Some lawyers are making money on this. But it is not fun litigation. Desks in the depositions have been arranged in ranks and rays."[126] For others, the delays, the millions of pages of documents, the shifting configurations of plaintiffs and defendants, the soaring legal expenses, and the legal intricacies all signaled that MDL 551 was a morass and a negotiated settlement a necessity. At the Supply System, Donald Mazur stated what others were feeling: "It's incomprehensible, when you look at the size and complexity of the litigation, that it will go full term in a trial. It is so massive that the legal system almost collapses. The matter begs for a solution."[127] In the spring of 1986, William Berls of Chemical Bank pointed out that a settlement funded by refinancing of bonds on the net-billed projects would be advantageous as long as interest rates remained low. He stressed the necessity for quick action: "It's an urgent situation. We have to move forward and

---

[125] Washington Public Power Supply System v. Pittsburgh-Des Moines Corporation, 876 F.2d 690 (9th Cir., 1988). See also Miller, *Energy Northwest*, 445–446.

[126] Gail Diane Cox, "Hunt for a Culprit Continues; WPPSS Round Two," *National Law Journal*, April 13, 1987, 1.

[127] Harriet King, "Efforts to Settle WPPSS Securities Litigation Are Gaining Headway," *Nucleonics Week*, June 12, 1986, 5.

settle quickly or the window could close." Even utility counsel Al Malanca expressed an interest in a negotiated agreement. "If there's a way to reach an amicable settlement, we'd certainly want to think about it."[128]

In September 1985, Judge Browning took a step toward a negotiated settlement. He appointed Junius Hoffman, a professor of securities law at the University of Arizona, to serve as settlement master. Browning instructed each party in the suit to name a representative to negotiate through Professor Hoffman and gave Hoffman the authority to hold settlement meetings with plaintiffs and defendants anywhere in the country. Hoffman would work in private; he and the negotiators were not to discuss the case with the press or other outsiders. He was also under instructions not to communicate with Judge Browning himself. The parties were to file briefs with the settlement master by January 6, 1986, outlining their positions on the litigation and potential settlements. The secretive nature of the negotiations Hoffman supervised makes it difficult to know how effective he was, but lawyers appreciated his professorial approach.[129]

At the end of 1986, Cyrus Noë, as close an observer as anyone, wrote rather plaintively, "I am entitled to hope that MDL 551 will settle sometime in 1987," but he doubted this could happen before late summer or fall, when the discovery process was slated to conclude. By spring 1987, he sounded even more pessimistic, fearing that no settlement could be reached without clarification of the standard of proof issues and that this couldn't occur before the trial in late 1988.[130] Four months later, however, on September 10, 1987, Junius Hoffman announced the first tentative settlement, between the investors and four major underwriting firms, who would pay a total of $92 million. There were uncertainties. The court would have to notify all bondholder plaintiffs of the agreement and Judge Browning would have to approve it. The underwriters refused to admit any wrongdoing. On the other side, some investors grumbled that this was not "nearly enough . . . a drop in the bucket." In December, Hoffman followed up the underwriter settlement with two smaller ones.[131]

[128] "Northwest Power Settlement Sought," *New York Times*, June 5, 1986.

[129] "WPPSS Judge Appoints Master to Settle Project 4, 5 Lawsuits," *The Bond Buyer*, September 20, 1985, 1; *Clearing Up*, September 20, 1985, 1; Harriet King, "Professor Has Key Role in Utility Default Case," *New York Times*, April 19, 1988.

[130] *Clearing Up*, December 31, 1986, 6; May 1, 1987, 7.

[131] Vicki Stamas, "WPPSS Suit Resolution May Be Long Way Off, Despite Settlement Offer," *The Bond Buyer*, September 11, 1987, 1; George Yacik, "Judge Agrees to Certify New Plaintiffs in WPPSS Suit," *The Bond Buyer*, September 15, 1987, 1; Vicki Stamas, "Bondholders Group in WPPSS Litigation Hires Muni Specialist to Assess Settlement," *The Bond Buyer*, September 16, 1987, 1.

In the first months of 1988, the settlement process stalled. At the beginning of April, Hoffman gathered all the parties and surprised them by directing plaintiffs to make a "global settlement offer." The gathering took place in private, and Hoffman ordered those in attendance to keep the discussion confidential, but a Washington utility manager reported that the plaintiffs had proposed that defendants pay about 47 cents on the dollar, approximately a billion dollars. The leak drew a resoundingly negative response from several Washington utility leaders, one saying that a total settlement of $300 million would be more reasonable.[132] Negotiators returned to hammering out agreements defendant by defendant.

Even before the meeting, the city of Seattle agreed to a $50 million payment. All but $6.8 million was to come from insurance policies the city held. The hefty settlement was ironic. Through its Energy 1990 study, Seattle had decided in the mid-1970s not to invest in Projects 4 and 5 (see chapter 3). However, as a member of the Supply System at the time, Seattle City Light representatives had participated in the System's decision-making, and signed the 1976 bond resolutions. The city had feared that its resources would be at risk if it went to trial; following the settlement with the underwriters, Seattle had become, as its city attorney commented, the "deepest pocket in the lawsuit. We are breathing easier tonight now that that exposure has been removed."[133]

Judge Browning had scheduled the trial for September 7, 1988, but in the spring and summer several defendant groups reached settlement accords with the investors and Chemical Bank. Nevertheless, as the opening date approached, the cases against major defendants remained unresolved. A week before the trial's opening, plaintiffs settled their claims against the Washington Public Power Supply System itself. The WPPSS pact was unique. It imposed no monetary cost upon the Supply System. Rather, it required the agency to work with plaintiffs. WPPSS would make its managers available to meet with bondholders' attorneys; they would testify in person in Tucson instead of from Richland via closed-circuit television. Crucially, the plaintiffs gained access to Harlan R. (Hank) Kosmata, a high-level Supply System executive who had worked closely with the Board of Directors and with the Participants' Committee.

[132] "Court Official Instructs Plaintiffs in WPPSS Case to Submit Offers," *The Bond Buyer*, April 4, 1988, 1; "WPPSS Plaintiffs May Be Willing to Settle Case for $1 Billion," *The Bond Buyer*, April 18, 1988, 1.

[133] "Seattle Agrees to $50 Million Settlement with Holders of Defaulted WPPSS Bonds," *The Bond Buyer*, April 13, 1988, 5; Jim Klahn, "Seattle Agrees to Pay $50 Million to WPPSS Bondholders," *Eugene Register-Guard*, April 13, 1988.

One of the plaintiffs' attorneys called the agreement "a real coup for us."[134]

Yet releasing the Supply System without its paying a cent outraged the Bond Investors Association, a group of angry investors led by Richard Lehmann. The Association vented its anger at Chemical Bank. The bond trustee had worked out the agreement with the Supply System secretly, "to keep you from learning the ugly truth... and how you have been sold short."[135] To make matters worse, the bondholders complained, the settlement would re-establish the Supply System's creditworthiness and allow it to refinance the net-billed Projects 1, 2, and 3, reducing its interest expense commitments by almost a billion dollars – with none of those savings going to the wronged investors. On the surface, this misconstrued the WPPSS settlement agreement, which said nothing about borrowing on the net-billed plants. However, the bondholders' bitterness rested on a real foundation. By agreeing to a settlement without payment, Chemical Bank had in effect conceded that it would never be able to tap into resources of the net-billed projects, implicitly recognizing the "Chinese Wall." With that concession, the Supply System could assure potential investors in the net-billed plants that their funds would not be diverted to pay off bondholders in the terminated plants. Thus, although no money changed hands, the Supply System accord was a big deal. (One of the agreement's benefits to WPPSS was that it could now close down its legal office in Tucson, staffed with thirty-five employees and costing a million dollars a month.[136])

Less than a week after the Supply System's agreement, but over six and a half years since it had terminated Projects 4 and 5, Multidistrict Litigation 551 formally opened. The cast of characters was sharply diminished; following the pretrial settlements, only 21 utilities and three "professional defendants" (architect-engineering firms EBASCO Services and United Engineers and Constructors and the Supply System's financial advisor, Blyth Eastman Paine Webber) remained as defendants. The hotel ballroom

---

[134] Vicky Stamas, "Appeals Court Clears Way for Star Witness to Testify on TV During WPPSS Trial," *The Bond Buyer*, August 29, 1988, 1; Stamas, "WPPSS to Settle with Holders of 4,5 Bonds; 25 Defendants Left," *The Bond Buyer*, September 1, 1988, 1; Stephen Labaton, "Partial Accord in Big Bond-Default Case," *New York Times*, September 1, 1988.

[135] Judge Browning quotes the BIA letter in accepting the settlement agreements on September 5, 1989, MDL 551 at 1402.

[136] "Bondholders abruptly let WPPSS off hook before trial," *Eugene Register-Guard*, September 1, 1988; Miller, *Energy Northwest*, 434, notes the Tucson office staffing.

site, no longer needed, had been abandoned for Tucson's federal court-house, and a local newspaper editor dismissed the trial, saying "It's a utility case. Yawn." However, it remained a very large utility case. The remaining utilities held nearly two-thirds of the project capability of the canceled plants. Over a hundred attorneys had found their way to Tucson. They were responsible for about 5 percent of all of the city's office space leasing for 1988.[137] Despite its slimmer dimensions, observers expected the trial could last a year. In its initial session, devoted to "housekeep-ing" measures, Judge Browning made it clear that the case would not be steamrolled to a conclusion. The court would recess for Jewish holidays in September, for a full week in October and another in November, and between December 5 and January 5, 1989.[138]

Considering the scale of the case, the jury voir dire went rapidly. A thou-sand citizens had filled out questionnaires the attorneys had prepared, and eighty-seven were questioned in a two-day period. Of the twenty selected, there were fourteen women and six men, with ages apparently ranging from twenties to seventies. Judge Browning planned to choose a smaller group by lot to deliberate a verdict at the trial's conclusion. The press reported that among those discharged were "a nun who opposed nuclear power, a man who couldn't stay awake as the judge gave instructions and a pregnant woman who said she was having contractions every 20 minutes."[139]

Following a Rosh Hashanah recess, attorneys for each side presented opening statements. Before the plaintiffs' counsel began, a squabble erupted when defendants complained that the plaintiffs' tables were blocking defendants' views of the jury and the witness box. (Lawyers continued to debate space allocations throughout the trial proceedings, even as settlements emptied out the courtroom.[140]) The plaintiffs' lawyers' statements were as assertive as their courtroom placement. Defendants had made "false statements...fraudulently...and negligently" in the official statements about WNP-4 and 5. They knew, contrary to assertions

---

[137] "Trial likely to be a business boost but a bore for Tucson," *Eugene Register-Guard*, September 4, 1988.

[138] Vicky Stamas, "Bondholders, Utilities Ready to Slug It Out at WPPSS Trial," *The Bond Buyer*, September 6, 1988, 1; Dennis Walters, "WPPSS Trial Opens on Anticlimac-tic Note with Lawyers Haggling over Housekeeping," *The Bond Buyer*, September 8, 1988, 1.

[139] "WPPSS fraud trial gets 20-member jury," *Eugene Register-Guard*, September 10, 1988.

[140] Dennis Walters, "Reporter's Notebook: Judge Delivers Last One-Liner in Court's Final Scene," *The Bond Buyer*, January 11, 1989, 4.

in the statements, that there was no need for the projects' power. More-over, while the defendants made the "promise . . . over and over" to pay investors, through the repeated take-or-pay clauses, at least some of them didn't intend to fulfill that pledge. "They intended first to litigate, [and] when push came to shove, they succeeded" in getting released from the commitment in the Washington State Supreme Court's 1983 decision.[141]

Albert Malanca, representing utilities joined together in the Washington Public Utilities Group, also presented a hard-hitting opening statement. In his statement, he ridiculed the plaintiffs' assertion that the participat-ing utilities he represented had "induced over 100 law firms to lie about the fact that my utilities had the authority to sign" the agreements and denied that his clients could have benefited from defrauding investors. Some officials in the utilities being sued had even bought bonds in the projects – an unlikely investment for anyone scheming to perpetrate a financial fraud. Malanca contended that the Bonneville Power Adminis-tration bore the responsibility for the failed nuclear undertakings. "It was Bonneville's plan," Malanca proclaimed, and BPA was a "central force in determining the need" for electricity by its role in demand forecasts.[142]

Despite the polarized opening statements, settlement talks accompanied trial maneuvers. On the last day of opening statements, the city of Rich-land, which had held nearly 2 percent of the projects' capability, agreed to pay $6.5 million to the plaintiffs. As Judge Browning later pointed out, Richland's presence as a defendant was potentially threatening to the plaintiffs' misrepresentation charges. Throughout, the city had asserted that the Participant's Agreement it signed was valid and had continued to make payments on the defaulted bonds into escrow, despite the 1983 decision removing its obligation to pay. Richland contradicted the plain-tiffs' portrait of utilities plotting to borrow money they did not plan to repay.[143] Conversely, this and other settlements put additional pressure on the remaining defendants to come to an agreement with the investors and Chemical Bank.

[141] "Plaintiffs' opening arguments kick WPPSS civil trial into motion," *The Oregonian*, September 15, 1988; Dennis Walters and Vicky Stamas, "WPPSS Defendants Seeking $2 Billion Damages for Misrepresentation of Facts," *The Bond Buyer*, September 15, 1988, 1.

[142] Dennis Walters and Vicky Stamas, "WPPSS Counsel, Bondholders Report Reaching Tentative Pact," *The Bond Buyer*, August 19, 1988, 1; Vicky Stamas and Dennis Walters, "Chemical Says It Will Represent Post-1983 Buyers of WPPSS Bonds," *The Bond Buyer*, September 16, 1988, 1; Dennis Walters, "Defendants Deny Defrauding Bond Buyers as Second Week of WPPSS Trial Concludes," *The Bond Buyer*, September 19, 1988, 1.

[143] MDL 551, 1405.

The settlement with the Supply System yielded the plaintiffs their opening witness, Hank Kosmata. Since Kosmata had been the agency's liaison to the Participants' Committee in the 1970s, the plaintiffs hoped to use his testimony to show that the Participants had covered up problems and uncertainties in the projects. However, Kosmata proved a prickly witness during his twenty-seven days on the stand. Presented with statements indicating that the Participants and the Supply System had cooperated to present a rosy image of WNP-4 and 5 to Wall Street and potential investors, Kosmata continually denied that the utilities had any intent to deceive. The notion of a conspiracy to defraud bondholders was, he contended, "ludicrous", "science fiction." Kosmata's stance angered plaintiffs' attorneys. One complained to Judge Browning that the witness was "making speeches [and] giving ... nonresponsive answers." Defense counsel joked that Kosmata sounded like a witness for their side. One afternoon, a utility lawyer gave him a thumbs-up sign as he completed his testimony.[144]

Contentious as Kosmata's testimony was, other developments during his time on the stand continued to shrink MDL 551's proportions. On November 14, fourteen participating utilities, holding almost half of the project capability in Projects 4 and 5 and including all twelve members of the Washington Public Utilities Group, settled. Among these were most of the largest stakeholders in the failed projects. In negotiations, *The Bond Buyer* reported, plaintiffs had demanded $226 million. Most of these utilities lacked insurance that would cover the settlement; they insisted they could only afford $181 million. To supplement their outlays, however, Bonneville Power Administration agreed to pay $35 million and the state of Washington, though never a defendant in MDL 551, promised to add $10 million more. For attorney Al Malanca's clients, it was a settlement "we can live with."[145]

[144] *The Bond Buyer*'s trial coverage reported on Kosmata's testimony. His characterizations of conspiracy charges are in Dennis Walters, "Trial Adjourns for Weeklong Recess on Note of Scorn Delivered by Witness," November 21, 1988, 4 and Walters, "Witness Says Idea of Utilities' Conspiracy to Mislead Investors Is 'Science Fiction'," December 2, 1988, 4. Attorney responses in Walters, "Focus of Proceedings Shifts to Role of Blyth Eastman as Financial Adviser," November 30, 1988, 1 and Walters, "Utility Officials Summoned for Meeting in Last-Ditch Effort to Reach Settlement," November 10, 1988, 4.

[145] Dennis Walters and Vicky Stamas, "14 Washington Utilities Agree to Settlement; Pact May Hasten End of Trial," *The Bond Buyer*, November 1, 1988, 1; (Associated Press), "14 Utilities Reach Accord in WPPSS Trial," *Los Angeles Times*, November 1, 1988. Judge Browning describes and analyzes the settlement at length in MDL 551, 1406–1417.

s "Consolidated Settlement" left only four defendants in the case: _mish PUD, the two architect-engineering firms who had attempted to build the plants, and the Supply System's financial advisor, Blyth Eastman. The Consolidated Settlement gave Snohomish the option of approving a payment on the same terms as the fourteen utilities, some $48.7 million, if it accepted the deal by November 10. After that, bondholders would raise their price by $10 million. Although Snohomish appeared headed toward rejecting the arrangement, late on November 10 it agreed.[146] By the end of November, two of the three remaining defendants – architect-engineers United Engineers and Constructors, Inc. and EBASCO Services – agreed to add at least $22 million to the settlement pot. When the trial adjourned at the end of Kosmata's testimony for a month-long holiday recess, only Blyth Eastman was left. The Supply System's investment advisor agreed on December 21, 1988, to pay $20 million, and the giant case had reached its resolution.

The Settlements: Consequences Without Truth

When some twenty attorneys and the jurors reassembled in January 1989, Judge Browning was in a genial mood. He joked that seeing the attorneys might strike fear in the hearts of the jurors but explained that the trial had ended with the final settlement. He introduced Special Master Junius Hoffman, telling the jurors, "He's the one who put you all out of work." But he also commented that there was still work to do. He would hold a hearing in the spring on the fairness of the settlements and on how the funds should be allocated.[147]

Summed up, the twenty-two individual and group settlement agreements provided $687 million for the plaintiffs. Judge Browning was later to call the figure "enormous"; it was larger than any previous recovery in a securities class action suit. Plaintiffs' attorneys, especially when they argued that the excellence of their work merited high legal fees, applauded their own efforts on behalf of the aggrieved investors. Defendants, on the

[146] Dennis Walters, "Snohomish Board Appears Unwilling to Accept Pending Settlement Offer," *The Bond Buyer*, November 11, 1988, 4; Walters, "WPPSS Enters Next Phase; Last Utility Holdout Accepts Settlement," *The Bond Buyer*, November 14, 1988, 1.
[147] Dennis Walters, "Reporter's Notebook: Lawyers Hold Reunion as Jurors Mull Their Role in Finance History," *The Bond Buyer*, January 10, 1989, 4; Walters, "Reporter's Notebook: Judge Delivers Last One-Liner in Court's Final Scene," *The Bond Buyer*, January 11, 1989, 4; (Associated Press), "Default Case Jury Let Go," *New York Times*, January 10, 1989.

other hand, could be relieved that their payments would not cripple the region's public utilities nor require steep rate increases. In some instances, insurance policies paid the bulk of defendants' settlements.[148]

Yet from the standpoint of the investors, the glass was more than two-thirds empty. Not only did the settlement fund amount to less than a third of the $2.25 billion of bonds sold for the terminated projects, but the bondholders had to wait nearly a decade before the payout. Allotments for Chemical Bank as bond trustee and for the squads of lawyers were going to cut further into the money available for widows, orphans, coupon clippers, and speculators alike. With those deductions, the bondholders could expect only about $590 million, or about 26 cents on the dollar invested. (Had all parties accepted the compromise that the Luce Panel proposed in 1983, the investors would have come out better off, bondholder activist Richard Lehmann pointed out with some bitterness.[149])

Bondholders complained about a host of problems with the settlements that ended MDL 551. Chemical Bank's failure to pursue the professional defendants let them escape too cheaply. The $35 million that Bonneville provided to the settlement fund was a pittance compared to the federal agency's access to funds and, even more, its responsibility for promoting and organizing the failed projects. The $10 million that Washington contributed to the Consolidated Settlement was equally galling; the state had escaped being named a defendant, despite its deep pockets and despite the fact that the State Auditor was responsible for reviewing and vouching for the bonds. Moreover, as noted earlier, this settlement reached into the Hoffer case. Chemical Bank's negotiators agreed to drop this case, where bondholders had sued Washington State for securities fraud, in exchange for the state's $10 million. Hoffer plaintiffs appealed but were eventually refused review by the U.S. Supreme Court.[150] That the Supply System

---

[148] Wood Dawson, bond counsel to the Supply System, was the exception to the rule of benign outcomes for defendants in MDL 551. When plaintiffs allowed it to settle for only half a million dollars, it revealed the venerable firm's weak financial base. In 1992, a larger firm, Hawkins Delafield and Wood, absorbed Wood Dawson. Commentators recalled that the suit had been a "pretty scary experience" for Wood Dawson. See Vicky Stamas, "Hawkins Delafield to Absorb Wood Dawson, WPPSS Bond Counsel," *The Bond Buyer*, September 8, 1992, 1.

[149] The value of the settlement is discussed in In Re: Washington Public Power Supply System Securities Litigation, 19 F.3d 1291 (9th Cir., 1994), 1303. Lehmann interview for observation about the Luce Panel.

[150] MDL 551, 1412–1417 for Washington state settlement; the 1992 Ninth Circuit Court of Appeals decision is at 955 F.2d 1268 (9th Cir., 1992); Dennis Walters, "WPPSS Bondholders on Verge of a Payout as the Supreme Court Declines to Hear Case," *The Bond*

itself could evade paying anything also rankled bondholders, especially since the agreement in effect cemented the barrier between the terminated plants and the agency's other assets. Some of the other settlements, too, drew derision or complaints from investors for failing to reflect either the financial resources or the perfidy of the defendants.[151]

In February 1989, on behalf of the WPPSS 4 & 5 Bondholders' Committee, Richard Lehmann's national group, the Bond Investors Association, sent a letter to all bondholders urging them to write to both Judge Browning and Chemical Bank objecting to the settlements. Chemical rejoined that the issues involved in the settlement were more complex than Lehmann's letter had claimed and announced its intention to send a report to all bondholders analyzing the settlements. Judge Browning waited until September to reply to the bondholders' accusations when he ratified the out-of-court agreements.[152]

The SEC Report: Truth Without Consequences

The negotiated agreements ending MDL 551 left the charges of fraud and misrepresentation legally unresolved. The settlements had allocated the monetary consequences of the default without a legal determination of the truth of the plaintiffs' accusations. This is not merely a reflection of the fact that the American legal system is adversarial rather than inquisitorial, seeking a verdict rather than an authoritative determination of reality. Despite the array of legal talent all parties had mustered and the painstaking efforts of Judges Bilby and Browning for more than five years, it is hard to conceive how this case could have been resolved in a jury trial. Bondholders who complained that settlements understated the culpability of certain defendants nevertheless must have known that the courtroom offered no guarantee of reaching the "right" outcome. The prospect of months of detailed discussion on recondite issues such as demand forecasting and municipal security disclosure requirements could not have augured well for a successful trial. The inevitable lengthy appeals of any

---

Buyer, November 3, 1992, 1; Paul M. Barrett, "Justices Clear Way for Bond Settlement to Be Distributed to WPSS Holders," *Wall Street Journal*, November 3, 1992.

[151] In his decision in MDL 551, Judge Browning reports and comments on the objections to specific agreements and to the overall settlement.

[152] MDL 551 at 1392–1395 describes and quotes from the BIA's letter and bondholder submissions in response to it; Dennis Walters, "Investors Association Writes to Bondholders to Marshal Opposition for WPPSS Settlement," *The Bond Buyer*, February 15, 1989, 1.

verdict would only have saddled the defendants with further uncertainty and forced the investors to wait even longer for any payoff. It took the onset of the trial (and the pretrial settlements) to press most parties to settle, but the likelihood that a full trial would resolve MDL 551 was surely slim.

If MDL 551 yielded consequences without truth, the September 1988 issuance of the Securities and Exchange Commission's *Staff Report on the Investigation in the Matter of Transactions in Washington Public Power Supply System Securities* offered truths without consequences. Initiated as default loomed in 1983 and expected to take two years, the study lasted over five. Rumors of its imminent release began in 1987, and investors grew restive waiting for publication. In early January 1988, SEC chairman David Ruder estimated the report would be done in four to eight weeks, but in March, with no apparent progress, Judge Browning wrote to the head of the SEC's enforcement division asking him to "expedite" the report. The next month Ruder announced that the report would come out some time in the summer. Chemical Bank's William Berls complained, "It's the largest municipal bond default in history, yet it doesn't seem to be a priority at the SEC. It baffles me. We're very frustrated." Meanwhile, the Supply System and those in the securities industry involved in shepherding the defaulted bonds to market hoped that the report would not be too damning.[153]

By June, strong hints emerged that the SEC staff would find fault with the securities industry's performance in marketing WPPSS bonds but would not recommend enforcement action against any of the parties. Instead, as it had in its 1979 investigation of the New York City fiscal crisis, it would recommend legislation requiring greater disclosure in the municipal bond market. The leaks provoked sharp criticism. *The Bond Buyer* quoted one trader as saying, "I can't imagine that after a five year study, using taxpayers' money, the government is going to wash its hands of WPPSS." In early September, as the MDL 551 trial was getting under-way, Ruder testified at a House Energy and Commerce Committee hearing and virtually conceded that the SEC would not take enforcement action; its budget was insufficient to take action. Representative John Dingell (D–Michigan), the committee's chairman, told Ruder that the WPPSS default was "rich in potential fraud" and required action, but the SEC

[153] "Federal Judge Asks SEC to Expedite Report of Findings on WPPSS Default," *The Bond Buyer*, March 3, 1988, 1; Vicky Stamas, "WPPSS Trustee Denounces SEC for Delaying Default Report," *The Bond Buyer*, April 25, 1988, 1.

head could only express hope that MDL 551 would offer some relief for the damage to investors.[154]

Although its language was guarded ("ponderous" according to *The Bond Buyer*) the *Staff Report* did present a damning picture of almost all parties. The Supply System itself "avoided the disclosure of negative developments." Bond counsel "failed to disclose ... that it was unwilling to opine on the validity of ten ... Participants' Agreements." Both bond counsel and special counsel to WPPSS presented issues concerning the legal implications of a failed project in a fashion "not as clear as those relying on the counsels' opinions might reasonably have assumed." All who worked on the official bond statements "did not seek negative information from the Supply System to the degree they might have." Underwriters "did not ... conduct the kind of investigation in [these] ... bonds that they perform in negotiated municipal bond sales and corporate securities offerings." Securities rating agencies "did not make independent verification of the information" they received from the Supply System and "were not aware of some undisclosed negative developments." Unit investment trusts bought project bonds "for their premium yield," without the kind of review of quality that investors deserved.[155]

In his cover letter to Representative Dingell, Ruder reiterated the grounds for the SEC's decision not to initiate enforcement action. Attempting to impose penalties would be complex, uncertain and costly. (In passing, Ruder noted reports that bond trustee attorneys had already earned $76 million, even before MDL 551 got underway.) MDL 551 and other court cases provided another forum for resolving the issues in the Supply System default. More affirmatively, the commission chairman asserted that regulatory measures would have a broader impact than enforcement actions confined to the Supply System.[156]

The SEC's decision not to take enforcement action against those involved with Supply System bonds brought relief in both the Pacific Northwest and New York. Al Malanca was glad that the utilities had not been singled out for criticism. A WPPSS spokesman expressed pleasure

---

[154] Vicky Stamas, "SEC Suggests WPPSS Report to Track Results of N.Y.C. Study," *The Bond Buyer*, May 26, 1988, 1; Craig T. Ferris, "It's Time SEC Stop the Games and Release Its WPPSS Report," *The Bond Buyer*, June 6, 1988, 1; Patrice Hill and Dennis Walters, "Ruder Asserts Lack of Funds Stymies SEC WPPSS Action," *The Bond Buyer*, September 12, 1988, 1.

[155] Comments in SEC, *Staff Report*, 373–376. "SEC Insists on More Muni Disclosure, but Takes No Action in WPPSS Case; Long-Awaited Study Criticizes All Involved in Bond Sales; Underwriters Are Relieved," *The Bond Buyer*, September 23, 1988, 1.

[156] Ruder's letter is at the front of SEC, *Staff Report*, no page number.

at the decision not to seek penalties. Learning that securities firms would escape punishment, one underwriter's employee simply responded, "That's wonderful," although representatives of other underwriters refused comment on the report. Supply System bondholders, on the other hand, had a negative response. Melvyn Weiss told *The Bond Buyer* that "The SEC does not know anything about the marketplace."[157]

The SEC followed the *Staff Report* with a statement on questions of municipal bond underwriters' obligations. For more than half a century, since the enactment of New Deal era securities legislation, municipal bonds had enjoyed exemption from many of the disclosure practices mandated for corporate securities issues. The Commission's restatement stressed that underwriters needed to engage in a "reasonable review" of bond issuers' disclosure documents for accuracy and completeness. In the case of WNP-4 and 5, where all but one of the bond issues were marketed on a competitive basis, underwriters had asserted that they were unable to carry out the kind of review they normally did in negotiated sales. The Commission granted that underwriters might have less information in competitive issues than in negotiated ones but nevertheless stated that they were required to perform a reasonable review. Since in many competitive issues there was actually little doubt about which securities firms would handle the underwriting, these underwriters would be held to a higher standard of review of the borrowers' claims.

To implement these principles, the SEC drafted a rule requiring issuers of municipal bonds worth $10 million or more to prepare a preliminary version of their official statement prior to the actual sale. This would go to potential underwriters who would then be required to send it to any person requesting it. Within two days of the sale, underwriters would have to acquire enough copies of a final official statement to make it available to all who asked for it. Although the Commission provided a ninety-day period for public comments on the rule, the expanded underwriter obligations took immediate effect; the head of SEC's market regulation division told securities firms they could ignore them only "at your peril."[158]

When the SEC published the final rule, 15c2–12, in the *Federal Register* on July 10, 1989, it followed the draft closely. The threshold offering size requiring a near-final offering statement had been lowered from

---

[157] "SEC Insists on More Muni Disclosure...," September 23, 1988, 1.

[158] The SEC interpretive statement and proposed rule are both reprinted in *The Bond Buyer*, September 26, 1988, 25; see also Vicky Stamas, "New Disclosure Rules Will Complicate Underwriting of Competitive Deals," *The Bond Buyer*, September 26, 1988, 1, and "SEC Disclosure Regulation Box Score," *The Bond Buyer*, October 3, 1988, 36.

$10 million to $1 million, but other technical and procedural regulatory provisions were looser than in the first version. Notably, commentators did not feel that the rule would prevent "another WPPSS." The SEC had taken action to prevent fraud, but, as the head of the Government Finance Officers Association put it, "What precipitated the WPPSS default was not an economic breakdown, but a court repudiation of debt." Nor did Rule 15c2–12 do anything to mandate disclosure in the secondary market for bonds, where they trade after the initial underwriting period.[159]

The municipal securities industry has adjusted to the requirements of Rule 15c2–12 (and strengthening amendments adopted in 1994) without major disruption. However, the consensus of legal scholars is that the disclosure provisions have done little good. A 1996 article found "no reported cases relating to the enforcement of Rule 15c2–12." In fact, between 1990 and 1994, the SEC reported "no significant fraud cases against municipal issuers or dealers." The title summed up the situation: "Same Problems – No Solutions."[160] These judgments, though severe, are less harsh than the verdict of securities law professor Joel Seligman on the SEC's handling of the Supply System case that precipitated the regulation: "The Securities and Exchange Commission's resolution of the Washington Public Power Supply System (WPPSS) investigation was the worst botch of a case in the agency's history."[161]

Conclusion

At first glance, the tortuous path toward resolution of the Washington Public Power Supply System default suggests the uniqueness of the Supply System's situation. The proximate cause of default – the Washington State Supreme Court's June 1983 decision in Chemical Bank v. WPPSS – was

---

[159] Jan Paschal, "Value of Proposed SEC Rule Debated," *The Bond Buyer*, March 31, 1989, 1; the text of the rule was published in *The Bond Buyer*, July 5, 1989, 22. The SEC final statement and rule are available at http://www.nabl.org/library/securities/rule15c212/34-26985.html, accessed August 9, 2005.

[160] Lisa M. Fairchild and Nan S. Ellis, "Rule 15c2–12: A Flawed Regulatory Framework Creates Pitfalls for Municipal Issuers," *Washington University Journal of Urban and Contemporary Law* 55, 1 (Winter 1999): 40; see also idem, "Municipal Bond Disclosure: Remaining Inadequacies of Mandatory Disclosure Under Rule 15c2–12," *Journal of Corporation Law* 23, 3 (Spring 1998): 439–467; Ann Judith Gellis, "Municipal Securities Market: Same Problems – No Solutions," *Delaware Journal of Corporate Law* 21 (1996): 427, 453.

[161] Joel Seligman, "The Washington Public Power Supply System Debacle," *Journal of Corporate Law* 14 (Summer 1989): 889.

itself a striking departure from the consensus of scholarly legal judgment on the validity of the Participants' Agreements. Invalidating the "take-or-pay" contracts surprised observers and outraged many who viewed them as tried and true financing devices. Just as the default had dwarfed other failures in the market for municipal bonds, the complexities of MDL 551 and related legal proceedings far surpassed garden-variety securities cases. In the twentieth century, it was rare, if not unheard of, for issues of state and local public finance to arouse the degree of public controversy that the WPPSS defaults had.

In another sense, however, the story of the Supply System default resounds with echoes of other trends in the political economy of late twentieth-century America. Certainly the failure of these reactor projects mirrored the history of nuclear power in the years after Three Mile Island. Growing public suspicion, tightened regulations, extended delays, and rising costs of construction and finance effectively closed off this escape route from the fossil fuel energy crisis of the early seventies. The Pacific Northwest had no monopoly on fractious energy politics. The financial history of WPPSS also paralleled national developments. Long-term finance with high interest rates was a challenge to projects around the country. Public borrowers relied more on revenue bonds than on general obligation issues. Since borrowers' general taxing power was no longer the guarantee of repayment, lenders' security was more and more enmeshed with the success or failure of a particular project. The special districts which were supplanting general-purpose governing bodies as issuers were, on the one hand, less likely to be accountable to an electorate for the policies they followed. On the other hand, the ratepayers' revolt that broke out in Washington was a sign that energy policy could mobilize both consumers and environmentalists around the country. The fact that the courts became the means of handling the conflicting interests in MDL 551 reminds us that litigation has become a crucial method of resolving issues of public concern. Thus, to consider the WPPSS default and its consequences a somewhat outlandish story in a remote corner of the United States is to miss its implications for American politics, public finance, and law.

7

Running Toward an Uncertain Future

"Those dudes are coming down," proclaimed William Counsil on January 13, 1995.[1] Counsil, Managing Director of the Washington Public Power Supply System, was announcing the demolition of Projects 1 and 3, the two reactors it had been preserving for eventual completion since the early 1980s. On May 13, 1994, the Supply System's Board of Directors voted nine to four to terminate the projects. Following a flurry of efforts to find buyers for the major components, Counsil offered to sell individual items from the projects' inventory. This "garage sale" approach continued on the organization's web site and in a store that the agency operated in Kennewick for several years. Since the Supply System was not organized to manage a retail operation, in late 1999 it switched from selling individual pieces to offering "a whole warehouse at a time" for scrap at a few cents on the dollar.[2]

The Supply System remained in the utility business, however. WNP-2 produced, at full capacity, about 1,200 megawatts of energy and the system also still managed the Packwood Lake hydroelectric station. The turbine generators that WPPSS had used to produce electricity from the Hanford Reservation's N-Reactor no longer operated, since the weapons plutonium production reactor had been shut down in 1987 for safety reasons. The closure progressed to termination of the reactor in 1992. With the shutdown of Portland General Electric's Trojan plant, also in 1992, WNP-2 became the only functioning nuclear plant in the Pacific Northwest, a quarter-century after the Hydro-Thermal Power Program

---

[1] "Agency to raze unfinished N-plants," *Eugene Register-Guard*, January 14, 1995.
[2] Chris Mulick, "Kennewick, Wash., Energy Supply Store to Discount Items for Quick Sale," Knight-Ridder/Tribune Business News, Lexis-Nexis, August 11, 1999, accessed March 13, 2000.

had envisioned twenty large new thermal generating stations, almost all nuclear, as the region's energy future.[3]

In 1998, the Supply System decided to shed the burden of a name that had long been a bitter joke. *Barron's Financial Dictionary* even had an entry for "Whoops" to refer to the agency.[4] On November 19, the Executive Board voted unanimously to change the organization's name to Energy Northwest. Changing signage and stationery would cost an estimated $140,000, but the Board felt it was well worth it to distance the organization from past failures. Ironically, an unforeseen complication intruded. Public agencies in Pacific Northwest states had been holding a series of conferences on weatherization programs for low-income households under the Energy Northwest rubric. They sued to block the WPPSS name change. "They don't want the stink of the old WPPSS rubbed off on their annual meeting," admitted Vic Parrish, CEO of the malodorous utility. When the groups reached an out-of-court settlement in June 1999, the Supply System agreed to pay $123,750 to fifty-five community service providers; in return, the weatherization group would find a new name and drop its objections to the Supply System's re-baptism.[5] Parrish contended that the name reflected new circumstances and a new organization. "We are not trying to run from our past, but run toward our future," he declared.[6] Regional, national, and even global forces were shaping that future.

The American Electricity Situation at Century's End

In the Pacific Northwest and beyond, by the late 1990s a new set of watchwords had supplanted both the construction-oriented supply side paradigm of mid-century and the soft-path scenarios of the 1970s and

---

[3] Spencer Heinz, "PGE gives up on Trojan," *Portland Oregonian*, January 5, 1993.

[4] Barron's entry cited in Michael B. Marois, "Don't Say 'Whoops': WPPSS Declares Name Change to Northeast [sic] Energy," *Bond Buyer*, November 23, 1998, 3. Ironically, even the leading trade paper covering the Supply System had to publish a correction to its story's headline the next day, noting that the new name was Energy Northwest.

[5] Marois, "Don't Say . . . "; Parrish quoted in Ola Kinnander, "Focused on the Bottom Line, WPPSS Cuts Costs," *Bond Buyer*, March 15, 1999, 1; settlement announced in "Energy Northwest Recent News Releases," www.wnp2.com/NEWS/recent.htm, accessed August 28, 1999.

[6] Marois, "Don't Say . . . " Parrish used the same line when, on April 27, 2000, Energy Northwest's Executive Board voted to change the name of WNP-2 to Columbia Generating Station. Energy Northwest, Corporate Information, News Release 00-04, April 27, 2000.

1980s. The key terms were deregulation and competition. A third word, less often heralded but perhaps even more crucial to understanding the new electrical energy regime, was gas. Falling natural gas prices from 1984 through the mid-1990s gave utilities an attractive answer to their supply and cost problems. Abundant gas production in the Rocky Mountains and Western Canada, surplus pipeline capacity, and intense competition in the natural gas industry proved a boon to utilities looking for more electric generating capacity. Gas prices at the wellhead had reached $2.66 per thousand cubic feet in 1984, thirteen times as high as they had been in 1972. Thereafter, the price declined almost without interruption to $1.55 per thousand cubic feet by 1995. Energy planners seemed persuaded that gas would remain cheap for years to come.[7]

Combined-cycle combustion turbines, fueled by natural gas, were proving to be the resource of choice for generation. The combustion turbines used reliable technologies that were still improving for further cost savings. Far less capital-intensive than the earlier era's giant coal and nuclear facilities, they could be brought on line quickly and shut down if demand did not warrant their operation. A typical gas turbine plant in the mid-1990s had a capacity about a fifth that of WNP-2, capital costs per kilowatt of $684 (WNP-2's had been about $3,000 in more valuable 1970s and 1980s dollars), and a construction lead time of only about four years, a third as long as WNP-2. Additionally, natural gas power plants produced substantially less pollution and fewer emissions linked to global warming than did coal, the fuel that supplied most of the nation's electricity.

During the years from 1991 to 1995, 57 percent of the new electrical energy resources in the Pacific Northwest came from gas turbines. Even though at the end of this period gas contributed only 7 percent of the region's generating capacity, this had surpassed nuclear's 5 percent share. A 2002 study by the Rand Corporation posited that almost all additions to regional supply would come from gas turbines. Nationally, the trend was even stronger; between 1992 and 2003, about 73 percent of new generation used natural gas.[8]

---

[7] U.S. Energy Information Agency, "U.S. Natural Gas Wellhead Price ($/Mcf)" at http://tonto.eia.doe.gov/dnav/ng/hist/n9190us3A.htm, accessed April 2, 2004; for the region's late-1990s optimism about gas, see Northwest Power Council, *Revised Fourth Northwest Conservation and Electric Power Plan*, July 1998, Document 98-22, online at http://www.nwcouncil.org/library/1998/98-22/Default.htm, accessed April 2, 2004. See especially Section 2-B, "Restructuring of the Natural Gas Industry", and Section 5-B, "Natural Gas Price Forecasts."

[8] Northwest Power Council, *Draft Fourth Northwest Conservation and Electric Power Plan: Northwest Power in Transition: Opportunities and Risks*, http://www.nwcouncil. org/library/1996/96-5/Default.htm, ch. 2, passim and especially pp. 2–8; ch. 5, p. 5;

During the 1980s, gradual decontrol of natural gas prices had accompanied their leveling off and decline. The favorable experience in that industry, along with the general hostility to government regulation during the Reagan presidency and beyond, provided a strong example for electrical power's own foray into deregulation. Deregulation marked a true revolution in thinking about electric power, long considered a natural monopoly industry. Since the early twentieth century, the policy alternatives had been public ownership or regulation of privately owned utilities, predominantly by state utility commissions. Large utility organizations, both private firms and public agencies, usually were integrated vertically, combining the functions of generation, transmission, and distribution. Except in very rare instances, neither retail customers nor the companies that distributed the electricity to them could shop among different suppliers for the electricity they used or delivered. In exchange for utility monopoly power, private corporations accepted both price regulation and the obligation to serve all customers within their territory.

By the late 1990s, wholesale electricity markets in the United States were increasingly deregulated. Utilities could now often buy electricity and related services (e.g., provision of reserves for peak demand times) from a variety of suppliers. This required technical and organizational access. The 1992 National Energy Policy Act broadened the access of "non-utility generators" of electricity to wholesale and retail markets and encouraged "wheeling," whereby transmission facility owners carry electricity generated by other firms over their power lines. Retail deregulation had also advanced. Many large end-use customers – industrial and commercial firms for the most part – now could choose their utility suppliers just as they can now select among different telecommunications providers. In a handful of communities, even residential customers can select their electrical service firm. California was in the vanguard of deregulation. A California electricity market opened in March 1998 with more than 200 marketers registered. Customers, including end-users of electricity, could buy power on hourly or daily contracts. If they needed more than their contracts provided, they could meet immediate needs on a real-time spot market where prices fluctuated on a second-by-second basis.

ch. 4, pp. 1–2. Hereafter cited as *Northwest Power in* Transition. Christopher G. Pernin et al., *Generating Electrical Power in the Pacific Northwest: Implications of Alternative Technologies* (Santa Monica, CA: RAND Science and Technology, 2002), 41–43. The report is online at http://www.rand.org/publications/MR/MR1604/MR1604.pdf, accessed June 17, 2005. For national trends, see U.S. Department of Energy, Energy Information Agency, *Electric Power Annual*, Table 2.1, p. 13, online at http://www.eia.doe.gov/cneaf/electricity/epa/epa.pdf#page=20, accessed June 17, 2005.

At the same time, the corporate world of electrical utilities, formerly one of the most staid sectors of American business, underwent an extreme makeover. Utility firms restyled themselves as energy service providers. They merged, expanded into new lines of business, looked abroad for corporate alliances and adopted new corporate identities, missions and images. The vicissitudes of Portland General Electric, Oregon's largest private electric utility, show how the industry's tortuous pathways wound through the Pacific Northwest. PGE, with a pedigree reaching back to 1889, served almost three-quarters of a million customers in Portland and environs. In the mid-1990s, Enron, the ballooning natural gas company, opted to enter the electricity business, hoping to profit as a trader in newly deregulated electric power markets. To become a player in electricity, it set its sights on acquiring a utility; PGE fit its specifications, and Enron completed a purchase deal in 1997. Enron President Jeffrey Skilling now could boast that the firm was the world's largest energy company, a claim repeated in his new vanity license plates, reading WLEC. Two years later, Enron was ready to try taking advantage of the untested scaffolding holding up the deregulated California market. It began to use its control of PGE's capacity to send energy to California in order to implement its schemes to "game" the Golden State's market. Its plans involved, among other manipulations, shipping power from California to Oregon in order to send it back southward at higher prices. Old-line Portland General Electric transmission employees resented their new masters' schemes – "the weirdest junk," as one put it.[9] Soon, Enron sought to unload its subsidiary, convinced it could carry out its energy trades without PGE. A deal to sell PGE fell through, and Portland General Electric became mired in the morass of scandal and mismanagement that led to Enron's bankruptcy in December 2001. Investor groups and a movement for a municipal takeover contended for the firm until, in April 2006, PGE became an independent company with shares going to Enron's creditors.

The competitive paradigm in electrical utilities, along with the abundance of low-cost fuel, brought about an overall decline in utility rates in the 1990s.[10] Coming after two decades of often-steep rate increases and alarms about shortages and the inevitable end of cheap electricity,

---

[9] Bethany McLean and Peter Elkind, *The Smartest Guys in the Room: The Amazing Rise and Scandalous Fall of Enron* (New York: Portfolio, 2003), passim, especially chapter 17, "Gaming California," provides an excellent account of PGE's fate while in Enron's grasp. "The weirdest junk" is quoted at 270.

[10] Corrected for inflation, the average retail price declined from 7.59 cents per kilowatt-hour in 1990 to 6.37 cents in 2000. U.S. Department of Energy, Energy Information Agency,

the new regime was celebrated as a triumph of market principles. Nevertheless, deregulation also posed serious, interrelated dangers. First, if deregulation promised lower prices for some, would that mean no benefits or higher prices for others? For residential consumers, who still rarely could reap savings from effective choice, allowing large commercial and industrial energy users to shop for the best deal from a variety of suppliers was no blessing. If factories, office complexes, and shopping centers could drive hard bargains with utilities, the utilities might seek to recoup their revenues lost in that sector from customers who had nowhere else to go.

In the industry, managers wrestled with "stranded costs." Some utilities had made large investments in generating facilities (many of them nuclear) that produced electricity at a cost above the deregulated market price. Estimates of the magnitude of these investments (as of the mid-1990s) ran as high as $100 billion.[11] In a regulated environment, public utility commissions normally would include those plants' costs in a utility's rate base, calculate a "fair rate of return" and set electricity prices to yield that return. Customers would pay for the firm's expensive power along with its lower cost output. With deregulation, if some energy users are able to switch away from firms with these expensive investments, either others less able to choose will have to pay the bill or utilities may be unable to collect enough revenue to pay their fixed-cost obligations. Thus stranded costs could become a problem both for equity and for the financial stability of the industry.

Another problematic aspect of electricity competition is its potential harm to conservation and environmental protection. Competition forces utilities to look for the cheapest electricity on the market, but if externalities are present these sources are not likely to reflect accurately all the costs, present and future, involved in producing them. If, hypothetically, a new fossil fuel plant could generate electricity for 3 cents per kilowatt-hour and a residential weatherization program could save each kilowatt-hour for

"Annual Energy Review 2002," Table 8.6, "Average Retail Prices of Electricity, 1960–2002," http://www.eia.doe.gov/emeu/aer/txt/ptb0806.html, accessed April 9, 2004.

[11] The figure comes from a trenchant report, Convergence Research, "On the Brink: Nuclear Debt, Subsidies, and the Future of the Bonneville Power Administration," Prepared for American Rivers, Trout Unlimited and Oregon Natural Resources Defense Council, September 1996. Available at http://www.converger.com/bpabrink/bpabrinklyz.htm, accessed April 9, 2004. Steven Weiss, "Competition Puts BPA at Crossroads – WPPSS Legacy Clouds Future of Power Agency," *Seattle Times*, June 16, 1996, is also informative. The U.S. Department of Energy's Energy Information Agency's 1996 study, *The Changing Structure of the Electric Power Industry: An Update*, December 1996, DOE/EIA-0562(96), cites a wide range of estimates but focuses on one of $87.8 billion (pp. 80–81).

4 cents, market pressures would induce the utility to invest in the generating facility, even if it did more than 1 cent per kilowatt-hour's damage
to the environment. Without regulatory incentives and pressures, utilities were tempted to cut back investments in conservation and renewable
energy sources. Many investments in conservation ("negawatts," to use
Amory Lovins' term) might be cost-effective – i.e., they would save a
kilowatt-hour for less than the marginal cost of producing it.[12] Indeed,
in 1996 the Northwest Power Council identified about 1,535 average
megawatts of cost-effective conservation measures available to the Pacific
Northwest (almost twice as much as the WNP-2 plant was producing).
However, there was no assurance that the market would in practice lead
either consumers or utilities to invest in these measures.[13]

Regional Developments

In the Northwest, energy surplus had been a defining feature of the utility
environment from the early 1980s through the early 1990s. Efforts to sell
excess power capacity and energy outside the region, especially to California, were generally unsuccessful. Despite energy abundance, environmentally sensitive Northwesterners involved in energy policy did press for
demand-side management, conservation and renewable resource development. The Northwest Power Council was usually receptive to these
approaches. Even Bonneville itself, previously quite impervious to such
notions, administered innovative programs in home weatherization, commercial lighting redesign and new, conservation-oriented building codes.
Conservation measures between 1980 and 1995 saved almost a thousand
average megawatts, about the effective output of a WNP-2 sized nuclear
plant. (By 2005, the Council reported nearly 2,500 average megawatts of
conservation achieved since 1980.[14])

These developments were less likely to grab headlines than the earlier
WPPSS crisis. Energy policy had become a lesser political concern. The
Supply System itself was a secondary element in regional energy supply;
it contributed about 9 percent of Bonneville's firm load. Most customers
of utilities buying power from Bonneville were unaware that, as late as

---

[12] For Lovins' approach, see, e.g., Joseph J. Romm and Amory B. Lovins, "Fueling a
Competitive Economy," *Foreign Affairs*, 71, (Winter 1992): 46–62 and Jon R. Luoma,
"Generate 'Nega-Watts,' Says Fossil Fuel Foe," *New York Times*, April 20, 1993.

[13] *Northwest Power in Transition*, p. I-5.

[14] Northwest Power Council, *Northwest Power in Transition*, ch. 4, pp. 1–2; idem, *The
Fifth Northwest Electric Power and Conservation Plan*, Document 2005–07, 35.

TABLE 7.1. *Debt and debt service for net-billed projects, 2007–2024*

	Columbia generating station (WNP-2)	WNP-1 and WNP-3 (terminated 1994)	Principal repayment	Interest payment
2007–2011	$1131.6	$1708.3	46.0%	54.0%
2012–2016	989.7	2501.4	69.7	30.3
2017–2021	943.4	1145.3	87.1	12.9
2022–2024	346.4	–	90.8	9.2
Total*	$3451.8	$5423.7	69.6%**	30.4%**

* Totals include interest balances equaling $118.3 million as of June 30, 2006.
** Principal and interest shares of total debt service reflect accounting adjustments for certain compound interest bonds.
Columbia Generating Station debt service completed 2024.
WNP-1 debt service completed 2017.
WNP-3 debt service completed 2018.
Source: Adapted from Energy Northwest, *2006 Annual Report*, 53, http://www.energy-northwest.com/downloads/FY2006ARFinancials.pdf, accessed January 12, 2007. Dollar figures in millions of dollars.

1999, perhaps 15 percent of their monthly electric bill was going to repay bond principal and interest not only for WNP-2 but for the other net-billed plants WNP-1 and 3, which were terminated in 1994.[15] That figure declined in later years with refinances at more favorable interest rates, but the cost has remained a substantial burden to Bonneville and its customers, as Table 7.1 indicates. Indeed the bonds will not be fully paid off until the year 2024. Between fiscal years 2007 and 2024, outlays will total over $8.8 billion.

As measured by press coverage, the ongoing work of the Supply System, after the 1984 opening of WNP-2, was barely visible. Operating reactors troubled anti-nuclear activists, but their efforts focused on PGE's Trojan plant. Repeated Oregon ballot measures to force the reactor to close for lack of a comprehensive program for radioactive waste disposal had failed, but in August 1992, with a new vote looming, PGE announced that it would shut the plant permanently in 1996. Cracks in metal tubing in the steam generators would require costly replacements that the utility did not wish to make. Days after anti-nuclear forces prevailed at the polls

---

[15] In fiscal 1999, Bonneville spent 24.9 percent of its total operating revenue ($651 million out of $2,618 million) on nonfederal debt service, almost all of it related to the net-billed projects. Since Bonneville's wholesale preference rate was approximately 60 percent of the average retail rate, we can calculate that roughly 15 percent of a typical user's electric bill went for WPPSS debt. See Bonneville Power Administration, *Annual Report*, 1999, p. 12 for figures on debt service and revenues.

that November, an accident forced the reactor to shut down. In January
1993 the utility stated that it would not restart it again.[16]

An emerging salmon crisis made energy politics reappear in new forms
in the mid-1990s Pacific Northwest. There were controversies about
the causes of sharp declines in the numbers of anadromous fish on the
Columbia and Snake Rivers and their tributaries, but most agreed that the
hydroelectric dam system kept juvenile salmon and steelhead from moving downstream and successfully adapting to their Pacific Ocean habitats.
One proposal was to breach four large dams on the Snake to ease passage to and from inland spawning sites. The idea infuriated Bonneville's
Direct Service Industry customers, irrigators and other large energy users,
as well as many utility managers, but what had first seemed like a radical
environmentalist fantasy earned serious consideration from policymakers by the late 1990s.[17] Although the dams are unlikely to come down,
the salmon crisis is almost certain to constrain hydropower supply in the
region. In December 2000, in the waning days of the Clinton administration, a federal "Basinwide Salmon Recovery Strategy" rejected dam
breaching but called for coordinated measures of "restoring habitat, lim-
iting harvest, reforming hatchery operations, and reducing the impacts of
hydropower."[18] Under George W. Bush, the federal government in 2004
propounded a less aggressive – but still expensive, at some $600 mil-
lion per year – plan for salmon recovery, but the following spring U.S.
District Court Judge James Redden ruled the administration's proposal
inadequate.[19]

---

[16] Paul Koberstein, "Trojan to Close in 1996," *Portland Oregonian*, August 11,
1992; Spencer Heinz, "PGE Gives Up on Trojan," *Portland Oregonian*, January 5,
1993.

[17] Several works provide valuable background on the Columbia River salmon situation.
They include: Joseph Cone, *A Common Fate: Endangered Salmon and the People of the
Pacific Northwest* (Corvallis, OR: Oregon State University Press, 1996); Cone and Sandy
Ridlington, eds., *The Northwest Salmon Crisis: A Documentary History* (Corvallis, OR:
Oregon State University Press, 1996); Blaine Harden, *A River Lost* (New York: W.W.
Norton, 1996); William Dietrich, *Northwest Passage: The Great Columbia River* (New
York: Simon & Schuster, 1995); Richard White, *The Organic Machine* (New York: Hill &
Wang, 1995); Jim Lichatowich, *Salmon Without Rivers* (Washington, DC: Island Press,
1999); Joseph E. Taylor, *Making Salmon: An Environmental History of the Northwest
Fisheries Crisis* (Seattle: University of Washington Press, 1999).

[18] "A Coordinated Federal Strategy for the Recovery of the Columbia-Snake River Basin
Salmon," December 2000. Online at http://www.salmonrecovery.gov/strategy.shtml,
accessed June 15, 2005.

[19] See, for example, Joe Rojas-Burke, "Judge Rips Federal Salmon Plan," *The Oregonian*,
May 27, 2005.

The drastic changes enacted and foreshadowed by deregulation in the electrical energy industry were reflected in the Pacific Northwest and in WPPSS and Energy Northwest itself. Bonneville's presence and the financial and political travails it faced in the mid-1990s shaped the region's response. At the heart of the matter was the fact that by then, Bonneville was no longer consistently the region's lowest-cost power supplier. Historically, disputes about the BPA had focused on which customers would get the limited supply of cheap federal hydropower that Bonneville marketed. More than half a century of controversies about the public preference clause, about Direct Service Industry rates, about interconnections and sales outside the region, and about access for residential customers of the Northwest's private utilities had all in effect been quarrels about how Bonneville's low-priced power would be allocated. Now, however, Bonneville's electricity was no longer a bargain for utilities and large industrial customers. With wholesale competition, the expansion of combined cycle gas turbine generation, and energy surpluses in California and the Southwest, Bonneville's customers could and in fact did take their business elsewhere.[20]

Bonneville's cost problems stemmed from several factors. In the first place, commentators generally agreed that the organization itself had not maintained high standards of efficiency. Blessed with cheap hydropower and shielded from competitive pressures, it had paid little attention to paring its operating costs. Moreover, as an agency born of New Deal era idealism, it retained at least a residue of commitment to public service.[21] However, Bonneville's main disadvantage in the 1990s resulted from the two regional cost burdens it bore – fish and wildlife protection, in particular salmon restoration expenses, and its obligation to pay off the debt on the three net-billed WPPSS projects. Bonneville was paying about half a billion dollars a year for the power of WNP-2 and the unfinished Projects 1 and 3. With an outlay of the same order of magnitude for salmon restoration, the agency was in deep trouble. In the mid-1990s, the wholesale spot market price for electricity from elsewhere in the West was between

---

[20] For example, in May 1995, one of the most successful aluminum smelters in the region, Northwest Aluminum, announced it would shift 40 percent of its electrical demand from BPA to investor-owned Washington Water Power. See James Marcus, "Aluminum's white knight: Brett Wilcox took a closed aluminum smelter in The Dalles and turned it into a money-maker," *Oregon Business*, 18, 10 (October 1995): 44.

[21] One should not exaggerate this difference. In his years as Bonneville Administrator, Peter Johnson tried to make the agency behave like a competitive enterprise despite its monopoly position. Johnson, "Why I Race...."

1 and 2 cents per kilowatt-hour. Bonneville had to sell its power at about 2.5 cents per firm kilowatt-hour in order to cover its fixed costs.[22]

Bonneville hoped to survive this period when it was being forced into competition with more efficient producers and maintain its customers while it restored its standing as a low-cost power marketer. One approach the agency contemplated was to become an aggressive marketer of electric power – joining the hunt for cheap energy and expanding the range of services it provided. A 1996 "Comprehensive Review of the Northwest Energy System," commissioned by the governors of Oregon, Washington, Idaho and, Montana,* took a more cautious approach. Contending that within a few years Bonneville's cost problems could be solved and it could once again become a low-cost power vendor, the Review's steering committee called for Bonneville to adopt a system of "subscriptions" – medium-to-long-term contracts with its customers to allocate its firm power at cost. Customers in effect would bet that, by committing to Bonneville while it was selling expensive electricity, they would reap future benefits from a supply of power at below-market prices. However, the amount subscribed would be approximately limited to the resources that Bonneville already possessed. If a customer wanted to serve substantial new loads, it would have to find its extra power elsewhere or work out a deal with BPA in which the utility, not the agency, assumed all of the risk of acquiring the new resource. Preference utilities, followed by current Direct Service Industry firms, would have priority in claiming subscriptions. The scheme provided incentives for customers to undertake long-term subscriptions. In short, the Comprehensive Review's strategy was to insulate Bonneville from the most intense competitive pressures by confining its role to serving current loads of existing customers and tying those customers to the BPA through subscriptions. In 1998, Bonneville itself adopted a Power Subscription Strategy, but, after objections from the DSIs, agreed to sell to them simultaneously with priority customers. Moreover, as BPA quietly mentioned in its subscription plan announcement, "The agency is planning to purchase some power to augment the existing system."[23]

---

[22] Cost estimates from Convergence Research, "On the Brink...."

[23] Northwest Power Council, *Comprehensive Review of the Northwest Energy System: Final Report*, Document Number 96 CR-26, December 12, 1996, online at http://www. nwcouncil.org/library/1996/cr96–26.htm, accessed April 9, 2004. U.S. Bonneville Power Administration, "Selling BPA Power in the 21st Century: Power Subscription Strategy," *Keeping Current*, December 1998, p. 3.

* Bonneville's service area includes the portion of Montana west of the Continental Divide.

As Bonneville worked to drive its costs down in the late 1990s, the market price of electricity was rising. Soon, customers saw Bonneville as the best supplier and locked in five- and ten-year contracts. When the agency totaled up its commitments in 2000, it had agreed to serve 3,300 average megawatts beyond its regular resource base. It now had to go to market to find most of that power, as prices started to soar. Giving the Direct Service Industries greater access to federal hydropower led to an anomalous, if not perverse, situation during the West Coast energy crisis of 2000–2001. Market prices for electricity soared; aluminum prices on the world market dipped. Aluminum companies in the Northwest shut down their production lines. Some of those with subscription contracts with Bonneville could sell their Federal System electricity back to the agency, which desperately needed the energy to serve its other customers. In December 2000, for example, Kaiser Aluminum sold electricity it had bought for $22.40 per megawatt-hour back to Bonneville at the astounding price of $500 per megawatt-hour.[24] In fiscal 2001, Bonneville spent nearly $2.3 billion on purchased power, eight times as much as it had two years earlier. For the most part, Bonneville clawed its way out of the crisis by paying its customers not to consume the electricity they had contracts to buy. This was cheaper than open market purchases in the rigged world of Enron, but it left Bonneville in a perilous financial state. The agency raised its wholesale power rates by an average of 43 percent for 2002 and faced another increase of perhaps 15 percent in 2003. However, by early 2004, BPA was favored with a good hydropower year and strong surplus sales to California. Bond-rating agencies responded with stronger evaluations of the agency's debt. Refunding of outstanding WPPSS bonds lowered interest costs and the agency appeared headed for recovery.[25]

The cost of recovery was high. The 2005 plan of the Northwest Power Council estimated the regional cost of the energy crisis at $6 billion in "increased power-purchase costs and foregone economic activity."

---

[24] Solveig Torvik, "Kaiser Aluminum Makes a Bundle by Reselling Power," *Seattle Post-Intelligencer*, December 11, 2000.

[25] For cost-cutting proposals, see Northwest Power Council, "Cost Review of the Federal Columbia River Power System Management Committee Recommendations," March 10, 1998, http://www.nwcouncil.org/library/1998/cr98-2.htm, accessed April 17, 2004; U.S. Department of the Interior, Bonneville Power Administration, "What Led to the Current BPA Financial Crisis? A BPA Report to the Region," April 18, 2003, http://www.bpa.gov/corporate/docs/2003/Report_to_Region.pdf, accessed April 16, 2004. For improvements in 2004, see BPA *Journal*, March 2004, p. 2 and April 2004, pp. 1–2.

Moreover, the Northwest's favored position among the nation's regions had shrunk drastically. In 1999, Washington's average retail electricity price was 60 percent of the national average. Four years later, it was 79 percent. For industrial power, the advantage had almost disappeared. Washington's price, 60 percent of the national average in 1999, was 93 percent of the 2003 figure. The position of Oregon and Idaho had also slipped markedly.[26]

One important reason for the convergence to the mean is that the turn of the century crisis closed most of the Northwest's aluminum smelters. They have not reopened and, according to the Northwest Council, there is an 80 percent probability that all will be closed during the next twenty years.[27] Although the Direct Service Industries had been a perpetual bête noir of the public utilities, their boosters had for more than half a century pointed out that they provided valuable flexibility for the Northwest's energy supply. They had taken cheap electricity to be sure, but usually with provisions allowing their power to be interrupted when the hydropower system faced low water conditions. Thus, the aluminum smelters had served as regional reserves for the rest of Bonneville's customers. The uncertainty that their closure introduced to the system posed a problem for Northwest energy planners.

Energy Northwest: The Supply System Reborn?

As a piece of the Northwest electrical puzzle, Energy Northwest faced many of the pressures that Bonneville was encountering. Its performance would help determine the region's energy fate in the twenty-first century. Three goals for the organization's performance were crucial: lowering finance costs, improving the productivity of the WNP-2 nuclear plant, and successfully recasting itself as a provider of a spectrum of energy services.

By the end of the 1990s, Energy Northwest had over a decade's experience in refinancing the high interest rate bonds it had sold for the three net-billed nuclear projects. As bond market interest rates declined, and as older bonds reached their call dates, the Supply System would replace the

---

[26] Relative prices calculated from U.S. Energy Information Agency, Department of Energy, "1990–2003 Average Price by State by Provider (EIA-861)," at www.eia.doe.gov/cneaf/ electricity/epa/average_price_state.xls, accessed June 21, 2005.

[27] *Fifth Northwest Plan*, Volume 2, p. 7–25. Online at http://www.nwcouncil.org/energy/ powerplan/plan/(07)%20Portfolio%20Analysis.pdf, accessed June 21, 2005.

costly borrowings with less burdensome obligations. Tax-exempt refunding had been permissible only for an operating plant, but in 1988 House Speaker Tom Foley, a Washington State Democrat, maneuvered a technical provision in tax legislation allowing WPPSS to refinance up to $2 billion for the mothballed projects.[28] Between 1989 and 1999, the organization carried out fourteen major refinancing operations, exchanging some $9 billion in high-interest obligations for lower-interest commitments. Overall, the average interest rate on its bonds declined from about 10 percent in 1989 to below 6 percent a decade later.[29]

A pariah on Wall Street since the early 1980s, the Supply System had managed to reappear frequently as a trustworthy borrower in the municipal bond market in 1989 and thereafter. For this turnaround, Bonneville's now-explicit guarantee that it would back the net-billed projects' bonds was largely responsible, since default by the federal agency was almost unthinkable. Bond rating agencies responded with solid credit ratings, well above the investment grade threshold. Large institutional investors bought the majority of the refunding issues. Their willingness to invest in the bonds showed market professionals' confidence in WPPSS's refinancing strategy and its organization. Speaking of the Supply System's debt managers, one banker commented, "They know their stuff and expect you to know it too."[30]

As an organization, by the new century Energy Northwest had come a long way since the Supply System's calamities of the early 1980s. The reformed Executive Board brought some experienced business people with more cosmopolitan perspectives into policymaking. Top managers' pay had become competitive with other large utilities – despite some sniping at executive compensation in 1993 when William Counsil's $250,000 salary topped all other public employees in the state.[31] Management reforms

[28] David Zigas, "Whoops: Investors May Let Bygones be Bygones," *Business Week*, September 4, 1989, 92, notes Rep. Foley's involvement. Many bonds are issued with call provisions allowing the issuer, after a certain date, to pay off the principal and any remaining interest due on the bond. When market interest rates are lower than the rate on the bonds, borrowers can save money by issuing new bonds. The proceeds pay off the called bonds.

[29] Michael B. Marois, "Steve Buck at Energy Northwest," *The Bond Buyer*, February 9, 1999, 12oa; Kinnander, "Focused on the Bottom Line."

[30] As of its 2003 *Annual Report*, Energy Northwest's bonds for the net-billed nuclear projects bore AA – ratings from Fitch and Standard and Poor's, Aa1 from Moody's. Online at http://www.energy-northwest.com/downloads/annual2003.pdf, p. 64, accessed April 19, 2004; Marois, "Steve Buck at Energy Northwest," 12oa.

[31] Joel Connelly, "Calls for Probe Follow WPPSS Pay Disclosure," *Seattle Post-Intelligencer*, March 5, 1993; Eric Pryne, "Trouble Still Dogs WPPSS," *Seattle Times*, March 11, 1993;

affected almost all aspects of the operation. A 1994 study had found that Supply System staffing levels were nearly 400 workers higher than at comparable plants. By downsizing to industry norms, the agency could save $20 to $30 million a year.[32] In the next five years, the agency went beyond this target, cutting staff by about 35 percent, from about 1,800 to somewhat over 1,100. Overtime expenses shrank by 86 percent in the same period. These and other measures at WNP-2 had reduced its average power cost from 3.34 cents per kilowatt-hour in 1995 to 2.26 cents in fiscal 1999, below both Bonneville's preference rate and the region's market price. Energy Northwest had ambitious plans for its nuclear plant. It invested in improved equipment to increase output capacity by about 130 megawatts, was moving toward a refueling cycle of twenty-four months instead of twelve, to decrease the plant's down time and its labor costs, and was preparing to ask the Nuclear Regulatory Commission to extend the reactor's operating license, due to expire in 2023, for two more decades.[33]

The third front of the organization's struggle to remake itself was diversification. Management realized that operating one nuclear plant, with a tainted history and a shaky performance record, would not ensure the Supply System's survival. However, the nuclear orientation and links to the military side of nuclear technology and politics remained, as the agency's flirtation with plutonium reveals. As far back as 1986, Washington politicians and businessmen associated with WPPSS had broached the idea of completing the mothballed WNP-1 plant on the Hanford Reservation to replace the Hanford N-Reactor. This would have maintained the Supply System's role in generating electricity from a plutonium-manufacturing facility's steam, just as it had for two decades with the N-Reactor. However, although the proposal looked enticing to such pro-nuclear stalwarts as Robert Ferguson, Charles Luce, and Senator Slade Gorton (R–Washington), it faced major technological obstacles as well as hostility from Oregon Senator Mark Hatfield and died quickly.[34]

Barbara A. Serrano, "Lowry Slashing Pay Rate of WPPSS Board Members," *Seattle Times*, August 17, 1993.

[32] Mark Holt, Congressional Research Service, "Comparison of WNP-2 Staff Levels with Industry Averages," Federal News Service, August 8, 1994, Lexis-Nexis, accessed October 4, 1998.

[33] Kinnander, "Focused on the Bottom Line" provides a good summary of Energy Northwest's efforts as of early 1999. The organization also provides information on its accomplishments on its website, http://www.energy-northwest.com. Figures here come from its annual report for 1999.

[34] See, for example, *Clearing Up*, July 25, 1986, 1; August 1, 1986, 1–2; August 29, 1986, 3.

Throughout the 1990s, the Supply System advanced proposals for expanding nuclear production beyond the ongoing operations of WNP-2. The most dramatic plan, however, did not come from within the agency. In the fall of 1993, a consortium led by people with links to the Supply System, including former Managing Director Ferguson, unveiled a scheme to complete Projects 1 and 3 and use plutonium from decommissioned nuclear weapons as fuel. These measures would be paired with the same actions in Russia in order to combine electricity production with bilateral disarmament. The consortium named the plan "Project Isaiah," after the Old Testament prophet who called upon nations to beat their swords into plowshares. They claimed to have an imposing $8 billion in financing lined up to complete construction of the plants. The finished projects would be deeded to the federal government, which would pay operating expenses, and provide a guaranteed amount of steam to generate electricity.

A National Academy of Sciences study released in January 1994 found Project Isaiah to be technically feasible. Converting weapons-grade plutonium into reactor fuel was an established manufacturing process. Hanford's Cold War legacy, indeed, gave it a head start for this. The Fast Flux Test Facility (FFTF), an experimental breeder reactor prototype, had been sitting without a mission since 1992. Auxiliary to the FFTF was an enormous Fuels and Materials Examination Facility, designed for remote handling of hazardous materials. Plutonium could be stored here while the mothballed projects were completed and then converted into reactor fuel. On the other hand, the National Academy was less enthusiastic about the economics of Project Isaiah. The costs of operating the reactors would outweigh the revenue generated, they calculated. Ultimately, paying for Isaiah "would simply amount to deficit financing by other means."[35]

Project Isaiah soon found a rival proposal, this from within the Supply System. This plan, announced in January 1994, de-emphasized the Russian connection but built on the principle that burning plutonium in civilian reactors was the best way to handle surplus nuclear weapons material. WPPSS wanted to complete WNP-1, on the politically welcoming Hanford site, but terminate western Washington's WNP-3 at Satsop. The Department of Energy would operate the plant. WNP-2 would also

---

[35] Quoted in Dave Airozo, "Seattle City Council Eyes Quitting WPPSS if Board Revives WNP-1, 3," *Nuclear Fuel*, February 28, 1994, 9. The National Academy of Sciences study, *Management and Disposition of Excess Weapons Plutonium*, is online, with the comments on Project Isaiah at http://www.nap.edu/openbook/0309050431/html/161.html, accessed August 2, 2005.

be transformed into a mixed-oxide (a combination of plutonium and ura-nium, known as MOX) fuel plant. Within a few months, under pressure from Bonneville, the Supply System announced it would terminate both WNP-1 and 3, but it continued to hold out hope for completion and con-version of WNP-1 to MOX operation for several years more. In April 1996, the aptly named Joe Burn, in charge of plutonium conversion for the Supply System, told a reporter that the organization wanted to finish WNP-1 as an MOX plant. Along with this and conversion of WNP-2, Burn also proposed that the Fast Flux Test Facility be reborn as a giant plutonium incinerator, burning the fuel without power generation and abandoning its research and development mission. "We'd only be adding heat to the desert," he contended.[36] When, in early 1997, Energy Secre-tary Hazel O'Leary announced that her department intended to dispose of some of its plutonium stockpile through MOX fuel fabrication, WPPSS quickly joined a consortium of engineering, consulting, and generating companies to design a proposal for a complete program of mixed-oxide fuel fabrication, plant conversion, and use.[37] Action elsewhere could not match the Supply System's eagerness for MOX. By mid-2006, Energy Northwest was out of the MOX picture; Congress had denied funding for construction of a mixed-oxide fuel plant at the NRC's Savannah River, South Carolina, site.[38]

As the Supply System's culture and the energy environment changed, it actively pursued diversification beyond nuclear energy. In the late 1990s, it planned to earn an additional $150 million in revenue by 2006 from activities other than its nuclear plant. This would represent nearly half its income. Management believed that competition would both threaten utilities tied exclusively to the lines of business they had pursued in a monopoly environment and provide profitable opportunities for those expanding their range. In a 1997 article, officials indicated an interest in everything from operating fiber-optic cable systems to raising salmon or perch on the Hanford Reservation in the cooling tower basins of Projects 1 and 4.[39]

---

[36] Don McManman, "Officials Like Team of WPPSS Plant, FFTF," *Tri-City Herald*, April 1, 1996.

[37] The Supply System's 1997 Annual Report hails Energy Secretary O'Leary's initiative and announces the organization's interest in mixed-oxide conversion.

[38] Mary O'Driscoll, "Oversight Hearing to Assess Zeroed-out MOX Fuel Program," *Energy and Environment Daily*, July 24, 2006.

[39] Don McManman, "WPPSS Plans New Projects, More Revenue," *Tri-City Herald*, April 12, 1997.

The results of diversification are considerably less impressive than the hopes of 1997. In fiscal year 2004, the Columbia Generating Station brought in 95 percent of the operating revenue of Energy Northwest.[40] Yet while the utility remained primarily in the business of generating nuclear energy, its final annual report of the twentieth century indicated the organization's broader focus. It envisaged involvement in almost all forms of power generation. It boasted that its small Packwood Lake hydro plant produced some of the electricity that Bonneville included when "green marketing" renewable, environmentally responsible energy resources. In addition to its gas turbine permits, Energy Northwest was looking at providing engineering and maintenance services to the Federal Columbia River hydro system, at offering instrument calibration services for environmental cleanup at the Hanford Reservation, at marketing financial, technical and business planning expertise to the regional energy industry, and at starting a "Center for Energy Innovation in Renewable and Distributed Generation Technologies." By 2002, the organization had put the Nine Canyon Wind Project, the nation's largest wind energy facility owned by a public utility, into commercial operation. It joined other agencies in the White Bluffs Solar Station, a small demonstration project. In 2005, it moved forward on plans to build two large (300 megawatt) plants that would be fueled by gasified coal. Gasification would result in a project with lower carbon emissions than a conventional coal plant, but approximately twice those from a natural gas-fired generator. There were also hopes that carbon dioxide emissions could be trapped and sequestered underground.[41]

Energy Northwest reached out beyond electricity generation as well. Rather than complete site restoration at Satsop, a difficult and costly venture, Energy Northwest had transferred its assets from the abandoned Projects 3 and 5 to a Satsop Redevelopment Project that tried to develop the area as an industrial park. Exploration of a similar venture was underway for the sites of Hanford Projects 1 and 4, through the so-called Benton

---

[40] Energy Northwest, *Annual Report 2004*. Online at http://www.energy-northwest.com/ downloads/FY2004AR.pdf, accessed June 24, 2005.

[41] Information on the renewable energy projects is on the Energy Northwest website: for Nine Canyon, see http://www.energy-northwest.com/gen/ninecanyon/index.html; on White Bluffs, http://www.energy-northwest.com/gen/whitebluffs/index.html, both accessed March 29, 2004. Coal gasification plans are outlined in Energy Northwest's News Release 05-16, "Energy Northwest Board Votes to Pursue IGCC Project," July 27, 2005, http://www.energy-northwest.com/news/index.php, accessed August 2, 2005. See also Chris Mulick, "Energy Northwest Mulls $1 Billion Power Plant Project," *Tri-City Herald*, July 11, 2005.

[County] Redevelopment Initiative. In 1998, the Supply System had joined with several public and private groups in the Tri-Cities area to start a "business incubator," the Applied Process Engineering Laboratory, in a former WPPSS warehouse in Richland. Although aimed primarily at nurturing start-up firms in energy and engineering fields, the APEL's first marketable product was a cosmetic, a lip moisturizer.[42]

Conclusion

A generation before Vic Parrish proclaimed that Energy Northwest was running toward its future, the leaders of the Washington Public Power Supply System ran toward a very different future. Although hindsight reveals it as folly, from their perspective the horizon was bright and the direction clearly marked. Cheap and abundant electrical energy had brought the Pacific Northwest into the modern era. The region's dynamism required continued development of power resources. Nuclear power would complement hydro, combining to assure a clean, low-cost energy supply. With the direction and financial backing of Bonneville, the Supply System expected to join the vanguard of the nation's march toward nuclear energy. The technical skills and politically congenial climate available in the Tri-Cities area near Hanford further reinforced the sense that WPPSS had a rendezvous with a nuclear destiny. In the mirror of self-analysis, the Supply System and its supporters saw the face of progress. To its advocates, public ownership of power supplies remained a progressive cause, challenging uncaring private monopolists and bringing power to all the people. The initial projects looked technically and financially viable. The electrical utility industry had a decades-long history of economies of scale and declining costs. The Supply System, like other special-purpose municipal borrowers sprouting up in the postwar era, faced generally receptive financial markets. The legal underpinnings of the arrangements felt secure. Finally, until the early 1970s, no broad popular movement against nuclear energy clouded the view of a nuclear future.

That future did not come to pass. Its disappearance reflects the transformation of much of American political and economic life in the last three decades of the twentieth century. Even in the late 1960s, the grounds for faith in continued economic growth based on cheap electricity and expanding energy supplies were growing shaky. American petroleum

---

[42] Energy Northwest Press Release 99-21, "Advanced Process Engineering Laboratory Boasts First Marketable Product," November 23, 1999.

output peaked in 1970; natural gas production reached its highest level a year later. The global geopolitics of petroleum has been a source of anxiety and danger ever since.

Initially, the oil shocks of the early 1970s convinced many that a switch to electricity would keep electrical demand growing rapidly. However, the electrical utility sector of the economy had its own severe problems. The costs of a coal-based electricity supply were also increasingly apparent, as problems like acid rain from smokestack emissions came to the fore. Initial hopes for nuclear energy stemmed in part from the "loss leader" strategy of electrical manufacturers in the turnkey era of the mid-1960s and the seductive promise that nuclear energy would be "too cheap to meter." Moreover, the expected economies of scale from ever-larger electrical generating plants were failing to materialize, regardless of the fuel they used. As we have seen, a very different strategy for nuclear development, perhaps based on a standardized plant design and a centralized government with the might and will to override popular opposition to siting them, might have led to different outcomes, but the American nuclear power industry a generation ago was probably doomed to disappoint its backers.

Demand forecasts based on methods and assumptions that had worked throughout the 1960s had also built confidence that the Northwest's commitment to nuclear expansion was prudent. The energy supply environment itself changed drastically in the early 1970s; with some lag, so did ways to predict the path of future demand. Forecasters became more adept with sophisticated econometric techniques. As energy prices rose, they recognized that this would curtail demand growth. Perhaps more importantly, forecasters became more aware of both the limits of long-range prediction and the ability of policymakers to affect energy demand. Forecasts could be exercises in "persuasive storytelling," not prophesies of a predetermined future.[43] These new developments were not unique to the Pacific Northwest, but some planners in the region were among the first to recognize that the boom years of mid-century were ending and that slower demand growth would be the norm. Unfortunately, the Supply System and Bonneville were not receptive to the new paradigm.

The shifting energy climate was probably in itself enough to derail full implementation of the Hydro-Thermal Power Program and the Supply System's role within it. Yet it might have fared better with another

---

[43] James A. Throgmorton, *Planning as Persuasive Storytelling* (Chicago: University of Chicago Press, 1996).

organization leading the way. In recent decades, critics have chastised American businesses for "hollowing out" – divesting themselves of key resources and losing organizational capabilities. The Supply System's problem was, in a sense, the converse; it tried to grow rapidly but did not attain the capacities needed to pull off the feat of building five large nuclear plants. WPPSS of course was not the only organization to fail to realize its goals. Enormous projects are enormously difficult. The litany of failed nuclear reactor projects, many undertaken by large, experienced organizations, demonstrates this, as do troubled projects ranging from the English Channel tunnel and EuroDisney to Boston's "Big Dig" highway project and efforts to site a permanent nuclear waste storage facility at Yucca Mountain, Nevada.

The changing financial environment in the 1970s and early 1980s initially made the scale of the Supply System's commitments appear feasible. In the tax-exempt bond market, shifts from municipalities to special purpose agencies as issuers and from general obligations to revenue bonds made the Supply System and its bonds look familiar to brokers and lenders. Although it was the largest municipal borrower in the late 1970s, the sums WPPSS was borrowing were only small fractions of the swelling volume of municipal issues: $24.3 billion in 1974, $48.6 billion in 1978, and $214.2 billion by 1985. Meanwhile, interest rates were growing to record levels: from 4.99 percent in 1973 to 10.88 percent in 1982 on high-grade municipal bonds. As construction schedules stretched out, the Supply System and others undertaking major projects found themselves capitalizing interest expenses – borrowing to pay the interest on earlier bond issues. Once this vicious circle was established, projects rarely could avoid huge cost overruns.

The complexities of large projects and the perilous financial environment were thus challenges facing a wide range of undertakings in the aftermath of the nation's post–World War II boom. The uneasy relationship of the peaceful atom to nuclear weaponry, however, was a complication that nuclear power faced alone. Early in the nuclear era, the State Department's 1946 Acheson-Lilienthal "Report on the International Control of Atomic Energy" put it bluntly: "The development of atomic energy for peaceful purposes and the development of atomic energy for bombs are in much of their course interchangeable and interdependent."[44] History and

---

[44] *A Report on the International Control of Atomic Energy.* Prepared for the Secretary of State's Committee on Atomic Energy. Department of State Publication 2498. (Washington, DC: U.S. Government Printing Office, 1946), 4.

geography implicated the Supply System especially deeply in those links. In some respects this was an asset for the Supply System. WPPSS began producing nuclear power by operating the turbine-generators attached to the plutonium-manufacturing N-Reactor at Hanford. Managerial personnel circulated between the Supply System and the military nuclear projects at Hanford and in the Tri-Cities area. The agency's Tri-Cities headquarters assured it of strong local support from a population steeped in a "nuclear culture."[45] However, the military containment could not withstand the forces that buffeted the WPPSS nuclear projects. The Nixon administration's sudden move to shut down the N-Reactor in early 1971, although reversed by intense lobbying, showed that the Supply System's inaugural nuclear project was a hostage to distant political and military forces at the AEC and the Pentagon (see chapter 2). Management experience in the institutions of the military-industrial complex might be the wrong background for dealing with market pressures, cost control measures, or Wall Street finance. The Supply System's location may have shielded it from local anti-nuclear protest but, if anything, it amplified the opposition that developed in western Washington and Oregon to the agency's massive nuclear commitment.

Anti-nuclear protest directed at the Supply System mirrored several of the key elements of the movement in other parts of the United States and abroad. A concern for "beauty, health and permanence," to borrow Samuel P. Hays's formulation of the core values of modern environmentalism, permeated the opposition. The movement combined direct action techniques (although never on the same scale as at Seabrook, New Hampshire) with electoral politics, taking advantage of Oregon's and Washington's liberal policies on ballot initiatives. It also deployed some of the counter-expertise that Brian Balogh points to as a hallmark of anti-nuclear politics nationally. Alternative demand forecasts provided damning evidence that the Supply System's projects were building excess capacity. Studies of the potential for conservation and renewable resources indicated that better choices were available. In the Pacific Northwest, opponents stressed the WPPSS projects' economic folly more than their potential safety and health dangers.[46]

---

[45] The phrase is from Paul Loeb, *Nuclear Culture: Living and Working in the World's Largest Atomic Complex* (paperback ed.; Philadelphia: New Society Publishers, 1986).

[46] Samuel P. Hays, *Beauty, Health and Permanence: Environmental Politics in the United States 1955–1985* (New York: Cambridge University Press, 1987); Brian Balogh, *Chain Reaction: Expert Debate and Public Participation in American Commercial Nuclear Power, 1945–1975* (New York: Cambridge University Press, 1991).

In sum, the vision that led the Washington Public Power Supply System into a rendezvous with nuclear power was blinkered and clouded. To realize that vision, the Supply System was the wrong agency, in the wrong place, and very much at the wrong historical time. Whether the reborn Energy Northwest will be able to run smoothly toward a different future remains to be seen. Whether a future of deregulation, privatization, and competition is the right one for the nation's energy policy should be a matter for democratic debate. Whether or not a new era of American nuclear energy becomes a reality, any future nuclear development must solve the technological, organizational, financial and political problems that plagued WPPSS and other failed projects around the country. And if the experience of the WPPSS holds lessons for today's dialogue, it is that meeting America's energy needs demands, above all, open-mindedness and flexibility, not dogmatic faith in a preordained destiny.

Index